1. Precursor

I think this is a situation in which the cosmological scientific community was so busy looking far and wide for answers about the design of the Universe something really obvious and simple got missed. I liken it to searching for the answer in an ordinary kitchen. Science looked in every nook and cranny and couldn't find it, making it necessary for someone to come along that could see the forest for the trees.

Yet it's so simple it's the stuff of high school chemistry lab using heat (metaphor of an oven) and cold (metaphor of a refrigerator) as catalysts for predictable reactions. In this case, extremely hot in the Big Bang (inflation) and near absolute zero cold in the compression of mass into a black hole, act as opposite catalysts for opposite states of mass with opposite sub-verses all in this one universe.

This Theory will lay out the multiple opposite properties of opposite sub-verses, the implications of which potentially answer many contradictions and mysteries, particularly regarding black holes. This will lead to an easy explanation for why universe expansion is accelerating, and later, an equally simple explanation for why the double slit tests produce particles and waves. But also how the 4th dimension is generated, along with the opposite, eliminated, having implications for oscillation, i.e. cycling of mass.

2. Opposite Sub-Verses

On the surface, the idea of extreme heat in the inflation acting as the catalyst for expansion may seem ludicrous from the standpoint of the event being deemed by science as an accident. Maybe it was and maybe it was not, but for now we will simply look at this idea and the opposite properties of these sub-verses to see where it leads, and if indeed it does satisfactorily answer previously vexing mysteries about the universe.

The way this got started was with a thought experiment. Now to understand how I came to be able to do thought experiments, you can skip to the last section, 25 of the book to find out or continue reading here, but essentially I describe myself as flexibly, partially autistic. To make a long story short I had a brush with autism at age three and a half, which was averted by creating a mental imaging game of rearranging furniture at a high rate of speed that caused random dyslexia. I postulate that the game released energy that would have otherwise caused severe autism. Then at age eight I willed myself to read straight across (instead of needing to intersperse occasional random words) which caused a split between the random dyslexic/partially autistic part and the more linear/day to day conventional thinking part. By flexible, I mean there is a way to direct concentration between a more or less autistic state. The autistic part can play with thoughts at a much deeper and larger spatial scale than the regular conscious mind.

My paternal grandfather worked on intelligence for the UK in WWII. My father was a mechanical engineer in SF, then in Silicon Valley, and later on nuclear bomb tech. in San Diego for the Government. My brother was an industrial designer for Atari having designing the top 1 & 3 producing pinball machines (at the time), with all of us partially autistic and geared towards design. I did not have drawing skills like they did and the Cad-Cam had not been invented yet (when I went to college), but I am predisposed to spatially work on difficult design problems so I used those skills to work on trying to figure out universe design. The theoretical ideas in this book are the result of that endeavor.

I set the parameters of the thought experiment mentioned in the paragraph above as the compression of a supernova into a super massive black hole with no other celestial bodies present, and for the test to be carried out several times for purposes of comparison. The results were always the same; not a singularity, but instead a very small volume with extreme density, a solid sphere.

Afterwards, I questioned whether the test results were accurate or simply the result of preconceived ideas as to what may occur, but I could not remember having had any set ideas on how the tests would come out. So I decided to run with it and see where it led. Instead of referring to this final compressed mass as a black sphere, I decided to keep the commonly referred name, black hole.

But then the question becomes, what happened on the way to compression towards the conventional expectation of a singularity that changed the results? After all if we look to the equation E=MC2, we understand that when the energy is gone, the mass is gone and that's a singularity, so something must have occurred just short of that eventuation, much like a baseball player getting tagged out just short of home plate.

After ruminating this over for some time, I decided to do an internet search for theories about the strong and weak nuclear forces. The reason why is because it made me wonder if they played a role in my thought experiment results. In that search, I found a 2004 Nobel peace prize winning Theory by three physics colleagues, Gross, Wilcek and Politzer, in which they theorized the **Strong force in the nucleus is dominant**. The inference is the weak force is recessive (or at least not dominant), and this made me wonder if it was conceivably possible for those roles to reverse (under sufficient force acting on the mass), i.e. switching to Weak force dominant/Strong force recessive as mass makes the final compression into a black hole, and if so, what would be the implications of such a switch.

I am not a particle physicist, so what a Weak force dominant compressed atom, particle or whatever it would best be described, is at this point unknown. However I would venture to say that after comparison of the different sub-verse properties it should be considered highly probable.

Without going any further, I decided to write tensors for this theoretical switching, always using W or S as dominant. In other words, using the letter to designate the force as it is recessive was not necessary. Also, it seemed best to use the plus or minus symbol (~) on the keyboard to indicate switching. Thus, they were written as follows:

S~W & W~S (Strong force dominant switching to Weak force dominant & Weak force dominant switching to Strong force dominant)

Of course then the question was, what could possibly act on these roles to cause them to switch? It later dawned on me that the Big Bang was extremely hot as in plasma hot and black holes are thought to be near absolute zero, -459.67F. Could extreme opposite temperatures be switching catalysts? Now one might wonder why the inflation would need to be so much hotter than the opposite cold temperature of compression into a black hole for these opposite switches to occur. However, keep in mind that **entropy** of mass precedes the point in time when near absolute zero cold acts as the switch catalyst, whereas in the inflation black hole mass must be fully energized from near zero. In other words most of the release of energy has already taken place in the mass, thus the cold temperature required for the switch does not need to be comparable to the heat of inflation.

Putting this together, in the inflation the tensor W~S is the switch that occurs via the catalyst of heat in the inflation, and the tensor S~W is the switch that occurs via the catalyst of near absolute cold in a black hole. Again the word that keeps coming up is **'opposites'**, but also it is the simplest way to cycle mass. Looking at the tensors above, you can see S switching to W leads to W switching to S, with the mass having come full circle, cycling from one expansion into the next. But that suggests a Big Crunch must occur instead of the universe expanding until it dims out, so questions still remain as to how that might occur and will be tackled later.

The implications of this took time to percolate, but at some point I realized that if this was correct, then opposite states of mass infers the proposition of opposite sub-verses. Then the question became what are their opposite properties and do they answer any long held mysteries about the universe?

First, I needed to define these opposite sub-verses, simply as S or W (dominant):

In S (which is the sub-verse we exist in now), gravity is a distortion of space-time and in W, gravity is a vacuum in the absence of space-time.

Opposite effects in opposite sub-verses, makes sense as we experience a comparatively mild form of gravitation as Earth's mass exerts a distortion in space-time (Einstein), versus what we indirectly observe in black holes, a vacuum gravitation that is 'pulling' in and eliminating surrounding space-time, expelling electromagnetism, converting gasses and denser mass into W dominant mass.

This vacuum results from mass returning to its primary state, W in which no space-time or EM exists. Recalling earlier in this Theory that black holes are not singularities, but rather having some volume and extreme density; a vacuum gravitation results from an incredible sum of mass existing in such a small volume. As the volume of the mass compresses in three directions or coordinates, X, Y & Z, gravitation transitions into a vacuum force. A vacuum gravitational force does not infer all matter is drawn into a black hole, as it can also be thrown far and wide.

The violent action that takes place when mass falls into a black hole is therefore not a surprise, as there is a collision between opposite sub-verses in which only one can retain its state, because opposite sub-verses cannot exist in the same place at the same time.

Secondarily, the energy in the mass is in conflict with the cold temperature W needs for the mass to remain in that state, expelling all electromagnetic energy, adding to the reaction so there are plenty of reasons for the violent reactions taking place.

This process of eliminating electromagnetic energy serves another purpose, as the black hole mass is now predisposed for renewal in the subsequent Big Bang to occur later by way of extreme heat, in which the universe will once again be ubiquitous with electromagnetism.

The black hole always wins these opposite sub-verse battles because energy is stripped from the mass as it encroaches on the event horizon, except in the inflation which will be explained later, because in that case the catalyst of extreme heat is delivered *inside the Black Hole core*. Once the mass reaches a certain threshold temperature, W can no longer maintain the mass in that state and switches to S for the inflation.

What about the 4th dimension fabric of space-time?

In the switch W~S, space-time is generated and in the switch S~W the opposite occurs; it is eliminated.

This would make sense as Earth causes a distortion in the fabric of space-time, with the opposite effect which would be a black hole residing in the absence of the 4th D., (to greater degrees of 4th dimension until there is full 4th at the event horizon). As a black hole pulls in gasses and denser forms of mass, it is eliminating 4th dimension. One could also state a black hole is recapturing what was disbursed in the inflation, like a closer on a door. The *door opening* (W~S) is 'the action' and the *closer* (4th dimension) is the equal and opposite 'reaction'. That reaction will not stop until all mass is Big Crunch complete. But whether it is pulling in matter or dormant, black holes exist in the absence of space-time.

Much speculation surrounds why light cannot escape black holes, but if the Theory of Opposite Sub-Verses is correct and black holes exist in the absence of space-time, then how can there be light?

Obviously light cannot travel without a medium (4th dimension), therefore it is technically inaccurate to say light cannot escape a black hole, as it is not possible for it to exist in that sub-verse because the electromagnetic spectrum and 4th dimension are non-existent in W.

This makes sense also from a standpoint of oscillation, as all space-time and EM would be eliminated by the conclusion of the Big Crunch, predisposed for the next inflation. Let's move on to another set of opposites; the extremes of saving the mass from destruction when either switch occurs:

In the inflation, switching to S saves the mass from annihilation, i.e.

expanding too far and in the switch to Weak force dominant saves the mass from compression into a singularity.

Besides the obvious opposites of S representing disbursed mass and W compacted mass, there is also sub-verse expression:

S leading to the periodic table of elements by way of compression via gravitation in stars, offering infinite possible combinations of design, i.e. infinite complexity, with W possessing uniformity.

That is to say, the form of matter making up black holes is the same as all other black holes. Therefore, S is complexity and W is uniformity. We also see uniformity just after the inflation with hydrogen. It is only in the compression of mass to denser elements later in an oscillation that complexity arises then ends up in uniformity again at the conclusion of the Big Crunch. You could think of the universe as originating in a small volume of uniformity, expanding out into space-time as a broad spectrum of complexity that can be expressed in infinite ways, then back into a small volume of uniformity.

The next question for comparison is what existed first, S or W?

The answer is W, which is referred to as the primary sub-verse and S is the secondary sub-verse in existence.

S sprang forth from W because complexity always arises out of uniformity (simplicity). Of course in an oscillating universe what came first may not matter that much (if Opposite Sub-Verses is proven correct), continually flowing from one into the next, but for academic purposes it's important to know their order of existence.

The last opposite comparison has to do with Universe shape. As many of us in science know, Dr. Saul Perlmutter (and his team) won a Nobel for the astronomical proof that universe expansion is accelerating. This put the scientific community in a spin because it had previously been presumed that gravitation would cause the universe to slow its expansion and fall back into a Big Crunch. With Saul's discovery there is a foreboding concern the universe will expand until it eventually dims out. Not a great reward for all of our intellectual skills in deciphering the universe so far.

However, while working on the various Opposite Sub-Verse comparisons it occurred to me that the primary sub-verse W starts out spherical, albeit comparatively small in volume as compared to sub-verse S. But S is flat, so is there a process occurring that will change universe shape back into a spherical shape in which a Big Crunch will occur? Well, if you look at these Opposite Sub-Verse comparisons, 4th dimension and electromagnetism being eliminated as the number of black holes and their size increases suggests an oscillating universe, as no space-time would exist before the next expansion. That suggests a cyclical process. The bi-line on the cover is, 'The Engine of Oscillation', therefore I suggest another proportional effect to be added on to the next section (3. Accelerating Universe Expansion), which if it can be proven will show how proportional events/effects in the oscillation of the universe change universe shape from flat to spherical. I wonder if Saul and his team through astronomical observation can detect that effect? That would certainly be a jubilant moment for all as we could all take solace in understanding the universe oscillates.

Note that sub-verse S can exist as universe shape changes in the direction of spherical, but cannot exist once the Big Crunch is complete. Therefore, S is designated as having a flat universe shape and W having a spherical shape, are the extremes, with combinations of both S & W existing in the middle range of those extremes. However, it is important to realize their inherent shape in the extremes because as we look to oscillation, the bridge between them should possess a process in which transition of one shape into the other can be detected astronomically (as explained in the next section, 3).

Opposite Sub-Verses comparison

Switch Tensors	W~S	S~W
Sub-Verses	S	W
Catalysts	Heat	Cold
Events	Inflation	Black Holes
4th Dimension	Generates	Eliminates
Electromgntsm	Ubiquitous	Non-existent
Gravitation	Distortion	Vacuum
Saves from	Annihilation	Singularity
Expression	Complexity	Uniformity
Existence	Secondary	Primary
Univ. Shape	Flat	Spherical

The magnitude of the impact it has from the quick comparison of these Opposite Sub-Verses is dramatic as we are stunned by how all the previously held contradictions about the universe simply melt away, realizing black holes are not operating with some physics we do not understand, but rather are doing exactly what we would expect in a universe with Opposite Sub-Verses.

With the opposite properties of opposite sub-verses explained so far, it is easy now to explain why universe expansion is accelerating:

3. Accelerating Universe Expansion

The 4th dimension is often described as the fabric of space-time. A predicted specific amount of lensing of light from astronomical observation proved its existence, thus we know from Einstein that the greater the mass the greater the distortion, the greater the gravitational attraction. So would that not suggest the 4th is composed in part by gravitation spread out over space and time? Does this not lead back to gravitation having been in one location in the first place prior to the inflation? Are black holes then not gravitation being brought back to its state of origin, compacted in a small cold volume? And would the 4th dimension as a whole have cohesiveness? And wouldn't the elimination of the 4th and EM by black holes (as theorized in the previous chapter) infer that increasing number and size of black holes is reducing 4th dimensional cohesiveness? And what would be the result of that progression?

Since the initial moment of the inflation it follows that due gravitation, at some point universe expansion should have started to slow. However astronomical observation of supernovae in the early universe has proven the expansion is in fact accelerating. This was followed by a Theory that there must be a dark energy that is a greater force than (gravitation) and 4th cohesion to cause the acceleration. Seems logical and at this point just about everybody in Cosmology has embraced that sequence of thoughts.

I ask that you bear with me as we take a time out to do a simple thought experiment, as follows:

Imagine a thick long sweater in a tub of water and I'm wearing a raincoat standing in an empty tub. I pull the sweater out of the water and put it on, and it falls by way of gravity, analogous to the force of inflation. As the sweater falls its cohesion should slow its fall, but what happens is people come up and randomly cut holes in the sweater, analogous to black holes. At some threshold of increasing number and size of openings in the sweater, it loses sufficient cohesion to resist the force of inflation and accelerates towards the floor, spreading outwards at the bottom edge.

It then occurred to me that this progression could be described best by way of simple proportion, as follows:

Increasing number and size of black holes

Is proportional to (is-P-to)

Reduced cohesiveness of the 4th dimension

Is-P-to

Reduced resistance to inflation

Is-P-to

Accelerating universe expansion

Is-P-to

Changing universe shape from flat to a Big Crunch, spherical

If before the Big Bang the universe existed as a Black Hole with some volume and extreme density, it would have been spinning. As the catalyst of extreme heat was delivered to the core (which will be delved into later), the mass above and below the equator would have had inherent trajectory deviant of flat, such that in the acceleration of the expansion predisposed the mass to eventually initiate an arc upwards or downwards (relative to the black hole equator), concluding in a Big Crunch. In other words the outer edges of the universe should be broadening, with the arc of trajectory accentuating until enough of the mass comes back into contact with gravitational attraction initiating, then completing a Big Crunch. Keep in mind that although a reduction in 4th dimensional cohesiveness is gradually increasing, there remains sufficient cohesion to keep the universe within its range as the shape dramatically changes. What starts out as a small arc to begin with will accentuate via 4th dimension cohesiveness.

The questions are; can this be proven via astronomical observation or by developing a computer program to play out the proportional events to see how universe shape changes and if so does it change from flat to spherical?

The above proportional events/effects are just five in the sequence of twelve describing a full oscillation. More aspects of the Theory as they relate to energy need to be delved into before delineating the full cycle in section 11.

As we can see, opposite sub-verses presents a simpler explanation than dark energy, which remains elusive. These five proportional sequences can be tested by compiling data to work equations to compare them to determine if simple proportionality can be confirmed to explain accelerating universe expansion. I do not have access to the data to make that determination, but it would be interesting if someone reading this accomplished the calculations and got them published.

At this point in the Theory, on the one hand I was emboldened only to realize there is a major hurdle. If you look back at the numerous opposite properties of opposite sub-verses, you may see the problem. It is in the form of a question:

Where does the thermal energy come from to act as the catalyst of extreme heat in the Black Hole to cause the switch of atomic force roles W~S to initiate the inflation?

Also, how does a switch of atomic force roles in the inflation generate the 4th dimension? Conceptually how do we even wrap our heads around how a dimension of those properties could be generated? As often happens in Science, one set of ideas simply leads to other even deeper questions.

In this case however, a simple conceptual construct of how the 4th dimension could possibly generate in the expansion was developed by testing different scenarios in randomly varied thought experiments. This would also need to account for where the thermal energy came from to cause the switch to initiate the inflation. Since everything so far had fit based on the underlying premise of **opposites**, I wondered if that idea would also help explain energy interaction.

Since mass expanded in the inflation, and in this Theory's case from a black hole, would there not need to be an equal and opposite form of matter that released its energy within the Black Hole? But also if mass has four forces, then wouldn't this opposite form of matter also have four forces? But since these forms of matter are opposite, it did not seem necessary for their respective forces to also be opposite, but instead parallel forces. In other words, just as there is Gravitation, there would be a similar force acting on this other form of matter. Just as there is Electromagnetism, this other form of matter would have its own spectrum of light, and so on.

Without debating over what this other form of matter might be, again I forged ahead to see where this would lead. Did it make sense in helping to understand what generated the 4th dimension?

If you do a thought experiment of the inflation using equal amounts of opposite forms of matter with parallel forces acting on them expanding into a new universe, it is easy to conceptually visualize that would generate an intermediary 4th dimension formed out those forces and their comingled differences expanding out over space and time.

I pondered this for weeks wondering if there was a mystery that could be solved to illustrate this mixing of forces that would expose equal amounts of opposite forms of matter. Then one day I was thinking about the mystery of the Double Slit tests that produce particles and waves, and I wondered if particles was representative of 3rd dimensional light, and waves representative of 4th dimensional light, a mix of 3rd and 5th dimensional spectrums of light. What light is in the 5th dimension would remain unknown, but maybe by knowing 3rd D. light is particles and 4th D. light is waves may help later to decipher what is 5th dimensional light. But most importantly it pointed to parallel forces acting on equal amounts of opposite forms of matter (that can hold and release energy).

This was starting to take shape, but what was disconcerting was the mystery of what this other form of matter might be. Had I taken a wrong turn somewhere in my thinking or was this possible? If it was possible, then **opposites** seemed so to be the underlying basis for universe design, accident or not.

Since Opposite Sub-Verses with its atomic force role switching mechanism might explain how mass can cycle, the inference would be we live in an oscillating universe. But cycling matter is only half of what would be necessary. The other half is *energy oscillation*. Whatever this other form of matter is, it would need to energetically work in concert with mass energy to provide an oscillating flow of energy from Big Bang to Big Crunch, into the next inflation. Since mass has entropy, from order (hydrogen) to disorder (black hole), this other form of matter would need to do the *opposite* (ectropy), from disorder to order so there is a differential flow of energy. Only through an *exchange of energy* could there be oscillation. This is obvious because if both had entropy or both had ectropy there would be no possibility for an exchange of energy for the next inflation.

This question lingered for quite some time until one day I read about Dr. Duncan MacDougall's controversial weight loss at death (WLD) tests in 1907, in which he theorized that if a person did have a soul it would have weight because its energy would mean it had mass. His tests results were inconclusive and later put into question by tests in Eastern Europe that garnered much lower WLD's.

Nonetheless, I had nowhere else to turn in the search for an opposite form of matter in another dimension. I thought about it some more and decided to run with the idea to see where it took the theory, much as I had done with other aspects of the theory related to matter oscillation previously explained. But I also did so because opposites had not failed this theory so far. If opposite sub-verses works as the engine of oscillation, then opposites should also work for the cycling of energy. Thus, the idea established itself as equal amounts of opposite forms of matter, mass and consciousness with four paired parallel forces. Now I would begin the process of analyzing their interaction. Also, at some juncture I would need to endeavor to write a series of predictions for attempts to prove the existence of this other form of matter.

4. One Thought Leads to Another

Opposite forms of matter, yes, but regarding energy some factors are parallel, such as:

'Energy is to mass, as thoughts are to consciousness.'

Meaning, thoughts are the energy of consciousness. You might respond to that by saying; 'No, thoughts are the energy of our physical minds', and that statement would also be correct. So how can both our consciousness and our minds have thoughts? They do so by way of a symbiotic, positive loop exchange in which both continually equalize to the same thought level. The mind and consciousness are fused from the standpoint of this life-long dynamic of thought level equalization that continually takes place. And in this manner consciousness utilizes life-forms to ascend to higher, more energetic thought levels, ectropy. As tantalizing as that dynamic may seem, it will need to be delved into greater detail later. For now we will continue:

If Einstein's equation E=MC2 (energy equals mass times the speed of light squared) represents the relationship between energy and mass, then a parallel equation must also exist to represent the value of thought in consciousness. Thoughts are to consciousness as energy is to mass, with the speed of light squared as the constant multiplier for both forms of matter. These parallel equations are therefore:

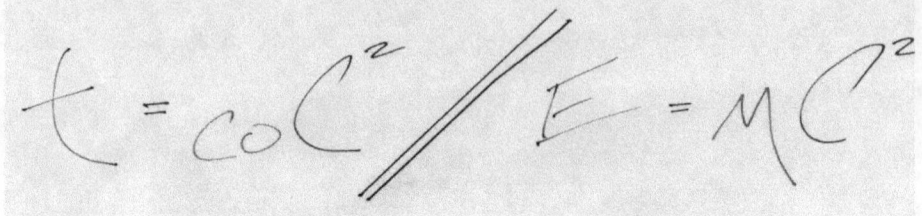

$$t = coC^2 \quad // \quad E = MC^2$$

- *thought = consciousness x the speed of light squared*

- *energy = mass x the speed of light squared*

The lower case t and co represent the thought level of lower consciousness in all life-forms including people. The two lines of separation indicate they are parallel equations. The constant, the speed of light squared, works for determining the energy in mass in the 3rd dimension, and since a parallel between mass and consciousness is being proposed here, it also applies for determining the amount of thought (energy) in consciousness in the 5th dimension.

Since the energy in mass is considered 'energy at rest', meaning the energy in the mass of the spoon you put in your mouth to consume soup will not release its energy from its atomic structure as you draw the liquid out of the valley of the flatware, then the 'thought at rest' in your consciousness, is similarly safe from releasing to say move a physical object in your path. The idea here is each one of us possess a certain thought level, and hypothetically, if that energy were released into for example a big boulder, it would either get hot and glow red, turn to lava, or possibly detonate. However, by the action of releasing your thought at rest into the mass, you would proportionately reduce your thought level with your eyes turning dull suggesting to observers you suddenly had the mind of a common farm animal, like a goat. This is why our thought level is held in check so it cannot be released, i.e. 'thought at rest'.

But what possible benefit could there be for us to possess a certain thought level at rest, that could conceivably be released into third dimensional mass under the right conditions? On the surface it may seem superfluous, unnecessary to the design of the Universe. But, let's be daring for a moment and suppose this notion has potential merit. What if all the mass and consciousness in the Universe, which have already been hypothesized to be equal in amount, are compressed into one Black Hole and one God, overlapped at the conclusion of the (previous) Big Crunch. Then the God consciousness (understanding the Universe cycle is complete) releases its 'Thought at rest' into the mass in the form of thermal energy. That energizes the mass into an expanding plasma, while reducing the God thought level into octillions of lowest level bits of consciousness disbursing with the mass into a New Universe? Tell you what let's not try to digest the whole Theory all at once. Let's partake of it at various moments, like wine and cheese tasting at a social event, and while we are at it, let's take a closer look at the details of how that might occur.

As we detour here, consider for a moment a Thesis entitled 'The Anthropic Principle', which was first suggested by astrophysicist, Brandon Carter of Cambridge University in 1973. It was an attempt to explain the observed fact that the fundamental constants of physics and chemistry just happen to allow for a Universe supporting life. A multitude of accidental parameters that just happen to fall into a range that supports the evolution of life. One interesting example is water, which is vital to life and unique amongst its molecules, is lighter in its solid state than its liquid form. If ice did not float, the oceans would freeze from the bottom up and the Earth would now be covered with ice. This is due to the unique properties of hydrogen. More examples given are; Gravitation is ten to the thirty ninth power times weaker than electromagnetism. If gravity had been ten to the thirty three times weaker, then stars would be a billion times less massive and would burn a million times faster. If the weak force had been slightly weaker, all hydrogen would have turned to helium and water would not have been possible, and life would not have propagated. A stronger nuclear strong force would have prevented the formation of protons yielding a Universe without atoms.

Another famous Scientist, Cosmologist, Sir Fred Hoyle, related one of his discoveries as having an anthropic basis. He observed that one particular nuclear reaction, the triple-alpha process, which generated carbon, would require the carbon nucleus to have a very specific energy for it to work. The large amount of carbon in the Universe, which makes it possible for carbon based life-forms to exist (e.g. humans), demonstrated that this reaction must work. He made a prediction of energy levels in the carbon nucleus that were later borne out by experiments. However, those energy levels were statistically unlikely. Hoyle was an atheist, however this discovery caused him to switch his beliefs to that of an agnostic deist, meaning it is more likely God exists than not. Here is his quote:

"Some super-calculating intellect must have designed the properties of the carbon atom otherwise the chance of my finding such an atom through the blind forces of nature would be utterly miniscule. A common sense interpretation of the facts suggests a super-intellect has monkeyed with physics, as well as with chemistry and biology, and that there are no blind forces worth speaking about in nature. The numbers one calculates from the facts seem to me so overwhelming as to put this conclusion almost beyond question."

If we are to presume a purely Scientific viewpoint of a physical only Universe, then why are all these fleshy things, flying, running and swimming all over this planet, and most likely on billions of other planets, by way of universally prodigious amounts of carbon? Additionally, why was the Universe designed to support their existence and evolution? Could it be that just as stars provide a medium with which mass compresses to denser, less energetic elements, life-forms act as a medium by which consciousness compresses to higher, more energetic thought levels? For this to be true, we must first reconcile that Evolution as proposed by Darwin works well enough for species evolution, so how would consciousness come into play? What is the relationship between them, and when did it begin?

To answer those questions, let's go back to our initial idea of all consciousness combined at the highest thought level transferring its thought at rest into the Black Hole mass to cause a Big Bang into a New Universe. Since energy is not free, i.e. it does not produce itself from the vacuum of space, there is a presumption that at minimum, there was a transfer of energy from a sending source to a receiving source, but in so doing the sending source must have been reduced by an equivalent amount of energy, and thus;

The 'Energy Exchange rule' (as defined in this Theory) states; the energy transferred into the mass, is proportional to the reduction in Thought level for the consciousness transferring it.'

A major portion or all of that consciousness would have been reduced to minimal thought, the lowest thought level possible and disbursed as a fine mist with all the mass, as octillions of barely perceptible thought level bits of consciousness.

At this point in the thought progression, I needed to understand how lowest level consciousness would have gotten started in life-forms. This idea came from asking a question to the autistic part of my mind; how did evolution initially get started between life-forms and consciousness?'

As I was starting to fall asleep a few days later, the autistic part presented an idea by asking, 'How about this?' and then played the following short scene: It was a view of a barren landscape with no foliage, insects or life, except for some microbes moving about in a shallow pool of water connected by a thin stream to a larger body of water. Tiny bits of light were dropping down and fusing with the microbes.

I thought about this and then realized;

'The first arising life-form(s) and minimal consciousness both possessed equal complexity to support the same thought level, and when fusing together initiated a symbiotic exchange, leading to higher thought levels and propagation of more evolved species.'

Water and electricity flow along the path of least resistance. Things that occur naturally do so with the least effort required. So when the first micro-organisms sprang forth in a pool of primordial soup, and simultaneously equally lowest level consciousness fused with them, no battle ensued to oust the consciousness. Rather it was simply accepted because it added survival value without any effort on the part of the life-form. This initial, simplistic connection occurred along the path of least resistance, initiating Evolution, defined in this Theory as:

'The symbiotic positive loop exchange between species evolution (Darwin), and individual consciousness evolution (Nevell), continually equalizing to the same thought level as DNA alters and consciousness compresses.'

Also, one could claim that the definition should start off by merely stating 'A symbiosis between...', however the interconnection between the two goes far beyond a simple symbiosis, which is generally referred to in Biology as a beneficial exchange between two species. In this case it is between the body and its consciousness, in which DNA alterations to support a higher thought level will immediately manifest by initiating a compression of the consciousness to the same thought level. The reverse is also true as any compression of the consciousness will initiate an alteration of the DNA to match thought levels. They are fused and synchronized, yet both forms of evolution are in a sense separate due to the time limited, transitory nature of their connection.

The inference is increases in individual consciousness occur over long periods of time in successive life-form fusing's, hitching a ride for the duration of those lives, and in the process elevating to higher thought levels. It is probably a case of numerous lives having some impact, interspersed by growth oriented lives accentuating the ascent.

Now, one could argue that what would seem like a rudimentary life-form to us, would not evolve by way of intellect and love. However, it is just that the parameters of consciousness compression are at a much lower level than humans, and as such we may not readily recognize them. For example; a certain fish species might hide in the sediment at the bottom of the sea, then launch a successful predatory attack from that position. Although instilled as instinctual behavior, it represents 'knowledge gained' sometime during that specie's evolution, which in turn is conveyed to its companion consciousness. It's not lofty, but none the less constitutes information about our Universe in the form of understanding the predatory value of a surprise attack. In this manner, every species has developed methods based on knowledge gained for survival. Conversely, certain species it attacks may develop the knowledge gained that open patches of sediment are not good places to linger, because there may be a surprise attack from underneath.

As far as love is concerned for these species, there is an equally rudimentary sense of connection with other consciousness, probably for the ones it mates with or gives birth to or sires. As an individual consciousness makes its way up the thought level ladder, existence in life-forms at each level is transitory, yet crucial for incremental thought level increases needed to ascend to the human level, and presumably to a genetically improved human version in the future.

'For every lowest level bit of consciousness, initially intellect and love are incrementally gained via simplistic, instinctual life-form existence, later transitioning to species with greater independent action, and over time and

a multitude of life-form existences, the result invariably will be an independent thinking, intellectual, loving consciousness.'

As we segue to human evolution, manipulating mass with opposable thumbs on limbs not used directly for mobility is a superior physical adaptation, facilitating an increased ability to gain knowledge. I state it in this manner, because there tends to be the perception that opposable thumbs are the end and be-all when in fact they are simply a means to an end. The 'means' being the physical adaptation, with the 'end' result being the knowledge gained, or at minimum an improved opportunity to gain knowledge.

It could even be put forth that a threshold of consciousness evolution had increased to the point where the physical adaptation to match a threshold thought level was expressed by an upright stance with opposable thumbs. This gets back to a 'what came first' routine, in which there is a presumption the adaptation occurred 'first', purely for reasons of survival adaptation by way of a changing environment or some such need, and by 'random' chance the adaptation just happened to lead to greater brain size. Conversely, it is being suggested here that it was probably pressure from ascending consciousness 'first' that pushed forward the adaptation. I know that notion will probably not sit well in scientific circles, however should the first and fourth consciousness predictions (in a later chapter) be proven correct via independent testing, supporting a whole new basis of understanding for consciousness ascension, this distinct possibility should be given its due consideration.

In a sense, does this idea not have its parallel with the Anthropic Principle? Are we not starting to recognize that all these accidental parameters capable of supporting life and the ascent of humankind, point with clarity in a particular direction? What will the threshold of 'knowledge gained' need to reach to understand a parallel in consciousness evolution?

The ability to gain knowledge is hard-wired into evolution, because peak level (god) consciousness has full intellect of how the Universe works, and that is the direction consciousness is moving. To illustrate how this can influence the speed of both forms of evolution, all we to need to do is look at, 'Lucy', the missing link fossil from Ethiopia of a human ancestor that lived 3.5 million years ago. She had the knees of an upright hominid, but the relatively small brain pan of a lower primate. So what happened after she stood up that changed the pace of our evolution? Once we stood up our hands were free from primary locomotion to manipulate mass, developing weapon making skills and painting cave art, providing 'knowledge gained', intellect. Once we could capture knowledge on parchment, information started to be stored for succeeding generations to draw from with our base of knowledge increasing rapidly, and with it our pace of evolution.

Now, one could claim that as a species our knowledge is great, yet as individuals it is comparatively fragmented. Meaning, each individual has only a minor viewpoint of all knowledge gained so far, which is correct. In a sense, it is the larger base of knowledge gained by our species that each individual enjoys in day to day life, benefitting from technological advances, medical knowledge, electro-mechanical devices, etc. Specialization has its advantages for society as a whole, but once a person understands that one half of consciousness thought level ascension is based on knowledge gained about all aspects of the Universe, then we have at least been made aware that;

'A greater understanding of a multitude of subjects is most advantageous in the pursuit of higher, more energetic thought levels.'

E.g., Thomas Jefferson made it a life-long pursuit of gaining knowledge in a multitude of disciplines, such as playing musical instruments, learning new languages, science and philosophy to name just a few. It is the person that has gained the most knowledge that best understands they do not know everything, and it is the person with the least knowledge that is most unaware of this fact. Ignorance may be bliss, but it is not prudent if one wants to ascend to higher thought levels.

Coupled with knowledge gained, is the non-physical half, which is the depth of connection with other consciousness, love, as the other major factor in our evolution towards higher thought levels. Those particularly inclined with an interest in Science, might on the surface be skeptical of the idea of love having anything to do with consciousness ascension, i.e. as integrally part of Universe design. However, love in this case is defined as depth of connection with other consciousness. This is due to the depth of connection that will need to occur at the time of the Big Crunch, for all consciousness to come together to think as One for the transfer of a thought at rest. At the stage of ascension so far achieved, consciousness has not combined, however, to release a thought at rest sufficient to initiate a Big Bang all consciousness must connect to think as One. This requires its own spectrum of consciousness ascension, evolving to eventually connect at a very deep level - let's just call it what it is, Love.

But just exactly how do we see this other half expressed in ourselves and other species symbiotic evolution? We see it expressed by way of social interaction. You will notice the most complex, intelligent species are the most social, as evidenced by their depth of connection with other consciousness. At the apex of social interaction are complex sounds forming languages communicated from verbal speech by Humans, but also by sounds made by dolphins and songs by whales. Until we are capable of compiling WLD's (weight loss at death - which will be delved into a future chapter) for these species, we can only roughly gauge by the degree of their brain structure complexity and the thought level each may possess. Chimpanzees, great apes, orangutans and bottlenose dolphins, killer whales are also very social species, reflecting a high thought level.

The incidental, yet integrally important side bar to our communication skills provides for deeper connection with other consciousness. "Romeo, Romeo, where art thou?" As the symbiotic positive loop exchange has progressed between our DNA and consciousness over millions of years, resulting in ever larger brain size, there continues to be:

'An even evolutionary emphasis in the two parameters for consciousness ascension as seen in our knowledge gained, Intellect, and depth of connection with other consciousness, Love.'

We can see this readily enough with humans, that we can intellectualize as well as we can love, and the two abilities often go hand in hand. As we learn to connect with each other at deeper levels, it is a prelude to the eventual connection of all gods and remaining lower consciousness, coalescing into one God, to think as One, for the transfer of Thought energy into the mass to initiate the next Big Bang.

Continuing with consciousness ascension; by observing grazing animals or even most predatory species, they tend to do what is needed for survival. We refrain from applying negative judgment towards their behavior. People mauled by bears or bitten by sharks most often do not blame the animal, suggesting they are innocent because they were simply responding to natural cues within their environment.

However, we do pass judgment on people within our own species for criminal behavior. This is based on the perception that they knew what they were doing was wrong. In fact, in many cases these crimes are pre-meditated. There is a saying that conveys this idea; 'A little bit of knowledge can be a dangerous thing'. Suggested here, is a transitional stage, or point where consciousness transcends from simply being subject to the basic instincts of the species it has fused with, to;

'Possessing a threshold thought level, supporting greater independent action, positive and, or negative.'

Accordingly, dolphins have been observed to conspire against a particular dolphin (a female for procreation) and a lower primate to rule in a ruthless manner. A person entrusted with leadership of a country might plunder that nation for personal fortune (e.g. Marcos' of the Philippines, Ceacescau of Romania or Qaddafi of Libya), or even start a war to gain more power and wealth (e.g. Alexander the Great, Ghengis Khan, Napoleon or Hitler). One outrageous example took place in the 1980's, when Romanian leader Nicolae Ceascescau appropriated one hundred city blocks (10x10) for the construction a new Capitol building complex. It was planned to be the largest, most ornate capitol building in the World. It cost so much money the Romanian people were pushed to the brink of starvation, causing tens of thousands of mal-nourished children to end up in orphanages. The Government reached a point of not being able to meet its financial obligations, yet the capitol was so immense it was nowhere near completion. A short time later Nicolae was overthrown in a revolution. At his trial he ranted that he was the best thing that ever happened to the Romanian people, and no one ever loved them more.

'This is a transitional period for consciousness that provides a great opportunity for independent action, yet also presents a dangerous period because of the odd and damaging occurrences that can transpire, individually and collectively, as the limits of this new found independence are tested.'

Absolute power corrupts absolutely, and the more money involved the greater the greed, are just two negative aspects of behavior providing a window into this transitional period of thought level ascension.

Following this new paradigm, as delineated and hopefully inspired by this Theory, presumably a new period will begin of greater understanding for our connection with all consciousness in its myriad life-form expressions, along with greater Intellect and love expressed between different individuals.

5. Out of Simplicity Arises Complexity

Let it be clear in presenting this new view of a greater evolution taking place in all life-forms, that Darwin's Theory is not diminished in any way by this new understanding. I see nothing in his Theory that can be construed as incorrect. It is just that two forms of evolution are taking place in the same medium, life-forms. It is as if we understood evolution by the layer Darwin provided, and now at this later point in time we add this new layer to fill out the big picture of this amazing interaction.

In a sense, they are both struggling for higher ground in order to gain as much advantage as possible against the pressures of survivability. One might suggest survivability is not important to consciousness because it is immortal, and therefore not confined to any specific physical lifetime. However, due to its fusing with a life-form having a primordial instinct to survive at any cost, consciousness is thoroughly convinced of the prime directive of survivability. This is the reference many suggest as 'the illusion of life', meaning each life lived seems like the only one we get, yet it is merely a temporary role we play in an ongoing series of curtain calls during our individualized ascending trek. Our memory for the previous curtain calls is unnecessary to our continued ascension, and in fact could potentially act as a distraction to the intended process.

Fact is ascending individual consciousness is not bound by existence in any particular species, so its contribution is transitory. Yet, to the extent that consciousness contributes to physical evolution is symbiotic, and therefore partially responsible for decisions determining a species destiny, for its continued evolution or eventual extinction. Intellectual arguments regarding how one form of evolution may affect the other could be endless, however suffice it to say both contribute equally to the ascent of the other, i.e. they share a positive loop exchange.

Along these lines, consider this; every mass extinction has been followed by a new set of organisms with a top tier species possessing a greater thought level than the previous periods' top species. Why? Because each period of physical evolution represents yet another period for consciousness ascension, and thus each subsequent period requires a higher physical order to support a higher (consciousness) thought level. Ascending consciousness cannot be proven by this idea however it does contribute support for the Theory.

Now a logical question on your part at this point might be; if there is a symbiotic positive loop exchange between specie evolution and individual consciousness evolution to ever higher thought levels in unison, then why has the crocodile not changed in one hundred and fifty million years? Evolution is a building process, and brain one as it is known is the R complex, the R standing for reptilian brain. It is the first brain layer at the top of the spine. The layer wrapped on top of it is brain two, the Mammalian brain. The third and final brain layer (so far) is the Neo-cortex, in primates and is most developed in Humans. Consequently the Crocodile is stuck at its stage of evolution. A profoundly perfect adaptation to its environment, to the extent it survived the great extinction of the dinosaurs, and provides a potentially great stopping off place for ascending individual consciousness. Yet in the big picture of this dual evolution taking place, it represents an evolutionary dead end. Conversely, Humans as a species are in the unique position of continually evolving, most likely because we are still catching up to the potential of the Neo-cortex in which we only use approximately fifteen percent.

As we can see from this Theory, there is an extrapolation of the Anthropic Principle to a greater degree than just stating that 'The Universe seems perfectly capable of supporting life', by stating;

'The Universe is designed to support life, for the purpose of individual consciousness evolution in the direction of all consciousness coalescing as a single entity, peak thought level God, to effect yet another Big Bang.'

By God, what is meant is a composite, a conglomerate like cells in an orange with each one of us having fully ascended, yet capable of thinking as part of One for the transfer of a thought at rest into the mass, and then being responsible for our individual ascension into the next expansion. Just as mass and consciousness are recycled to save on recreating those forms of matter, I also suspect individual consciousness recycles. The advantage is probably faster individual ascents in each subsequent expansion as experience hones our skills, as well as more creative and interesting Universe's. If on the other hand individuals only experienced one expansion to then be dissolved so to speak into a single consciousness at the conclusion of each Big Crunch, then each subsequent Universe expansion would become more predictable. By recycling individual consciousness, each subsequent expansion is like a flower that keeps getting more beautiful.

Thus, we can see from Carter's Anthropic Principle and Darwin's Theory of Evolution, each represents underlying foundations for the extrapolative assertions in the Theory of Opposite Sub-Verses. The Anthropic Principle providing a view into a Universe supportive of life-forms, and Darwin's Theory offering a view into evolving life-forms, both of which are partial views and supportive aspects of a Universe designed for individual consciousness evolution. In summation, we can say that all three are integral design aspects supportive of a single overall intended outcome.

Einstein who worked on a 'Theory of All' (describing how the Universe works), stated at one point, "When you do hear the right explanation, it will be simple, because out of simplicity arises complexity." And what could be a more simple beginning than lowest level consciousness fusing with earliest appearing micro-organisms to initiate a symbiotic positive loop exchange to ever higher thought levels for both, leading to the propagation of millions of species, including its apex specie, humans, with presumably a more evolved version in the future? Similarly, what more of a simple a beginning could there be than hydrogen converging to form stars, and in their hot cores and supernovae (end of stars life) compressing mass into the full spectrum of mass densities listed on the periodic table of contents, providing a multitude of elements for potentially infinite technological designs?

'Out of simple beginnings in our Universe, utilizing uniform building blocks, by way of compression, arise spectrums of infinite expression, then back to the uniformity and simplicity of one God overlapping one Black Hole.'

For some, it may seem like an objectionable transition that must have taken place to achieve our current thought level, having elevated to this point via a plethora of lower life forms. But realize that one cannot begin the journey at the top of the mountain, or even part way up the mountain, but rather it must begin at the base of primordial evolution. From another viewpoint, one could view it as a beautiful realization to have achieved such an amazingly demanding challenge, with such a diverse multitude of life-form perspectives having been experienced to eventuate at our current level. Sometimes I am grateful for my level, and when looking at lower life-forms as we drive along, like cows grazing in a field, I've said to my wife, Paula, "I sure am glad I'm not a cow. How boring would that be?" But of course the thought level of a consciousness in a cow is commensurate with grazing, just as the thought level of a Pine tree is commensurate with its seasonal functions.

'The beauty of bio-diversity is that all levels are available for ascension.'

By the same token, some of us may be attempting to streak to the finish line to ascend as soon as possible to the god level. However, keep in mind that for everything gained something is lost. After transitioning into a highest level god, what will be gained is a peak thought level, but what will be lost is the challenge of ascent. You see it is a very different state to be at the end of the line looking down with nowhere to go higher, just waiting for the Big Crunch than it is to be challenged as we are by each next step up. So I say enjoy cohabitating in a physical life-form, in which you as an individual have every opportunity to;

'Stop and smell the roses'.

6. The Consciousness Predictions

When Einstein's prediction of gravitational lensing of sunlight around the edge of the Moon during a solar eclipse was proven correct by astronomical observation, it suggested a wrinkle in the Universe. In other words, in a Universe with all things being equal, light should shine straight. The fact it did not under specific circumstances is the wrinkle that offered up the fourth dimension.

In a similar vein, by virtue of the following predictions what is being suggested is 'Weight Loss at Death' (WLD) is comparably equivalent to thought level, not physical size of an individual or the size of a species. In this manner, it will represent a wrinkle, because in a Universe with all things being equal, any measurable WLD should correlate to physical size. When upon comparison of a multitude of species it is understood it has everything to do with thought level, not size, it will be realized that a window into the 5th dimension has been opened for the first time in Human history.

Before coming to this idea of measuring WLD, I searched the internet for work or attempted experiments that had taken place that might help guide or provide a clue as to how the process might best be accomplished to somehow quantify consciousness. Then finally one night in late 2010, I noticed on a message board on the internet, a post someone had written in response to the question "Is there a soul?" This post asserted scientists had measured the weight of Human souls. Although this seemed highly unlikely, a search was initiated to find out if that was the case. To my great surprise, Dr. Duncan MacDougall had conducted tests on patients dying from Tuberculosis in 1907, to see if there was any weight loss at death. His assertion was that all mass has energy, and if consciousness (a person's soul) has energy, then there should be some measurable weight loss at death as it exits the body.

He conjectured that TB patients would work well for these experiments because it was easy to recognize when they were in their last hours of life and they remained still, which he figured would keep the balance bar steady at the moment of measurement. Upon news of a patient nearing death, he would roll their bed on to an industrial sized weight scale accurate to the gram, and maintain the balance bar at zero. He claimed there was a weight loss right at the moment of death that could not be confused with weight loss over time due to respiration (loss of moisture from the lungs from breathing). His tests were conducted on six patients, with four conclusive tests, the average WLD (weight loss at death) was 21 grams. In fact, there is a movie entitled '21 Grams', which unfortunately has nothing to do with his experiments, however at the very end of the movie this is what is said:

How many lives do we live? How many times do we die? They say we all lose 21 grams at the exact moment of our death. And how much fits into 21 grams? How much is lost? When do we lose 21 grams? How much goes with them? How much is gained? Twenty one grams is the weight of a stack of five nickels. It's the weight of a hummingbird, a chocolate bar. How much did 21 grams weigh?

Poetically inquisitive, particularly with comparisons of fifth dimensional soul/consciousness weight in third dimensional equivalents and it should be noted 21 grams was the average with fluctuations between different patients. Although the weight is not significant in relation to comparable forms of mass, it does represent the co quotient in the equation $t=coC2$, and therefore represents a considerable amount of 'thought at rest' energy. Keep in mind that when Einstein opened people's eyes to the equation $E=MC2$, it was suddenly understood just how much energy was in mass. Now open your mind to that same idea for consciousness.

Dr. MacDougall later conducted tests on fifteen dogs, and although the weight loss was not discernible with his equipment, he admitted the tests did not occur in the same manner as with people. Nonetheless, his conclusion was that dogs do not have souls (consciousness), but that was also his preconception before conducting the tests. Since a dog has a much lower thought level than a Human, it is my assertion his equipment was not sensitive enough to register a reading. How much higher a thought level is a person operating at than a dog? In any case, my estimate of what a dogs' WLD would be is less than one gram. Since his equipment was accurate to grams, it's likely none of the dogs WLD's were sufficient to tilt the scales. This is a perfect example of preconceived notions leading to and influencing inaccurate test conclusions.

During the early part of the 20th century when the good Dr.'s tests took place, it caused quite a sensation with the publication of his test results in many newspaper articles around the United States. And as you might surmise, especially in 1907, there were many liberties taken with the truth by the press. Understanding this all too well, one newspaper publication provided Dr. MacDougall with an opportunity to set the record straight about his experiments. At the conclusion of his statement was this quote;

"I am well aware that these few experiments do not prove the matter anymore than a few swallows make a summer, yet the results should at least provoke further experiments." Dr. Duncan MacDougall

The unified response from the scientific community of 1907 was his experiments provided nothing more than a mere curiosity. However, that seems an unfortunate conclusion to reach given the value of such experiments to potentially prove the existence of consciousness. I once asked a gold prospector in the gold country of California where to find gold. He said, "Gold is where you find it." Well, Dr. MacDougall's unfinished experiments just may be gold that has been untouched since 1907.

Since then other WLD tests have been conducted on sheep, goats, mice and in Eastern Europe human tests. All of them in my opinion have been inconclusive for two reasons: 1) they did conduct the test quickly enough. Any lag time and measuring a minute weight loss cannot occur. It must occur within a few seconds (but without causing vibration which will change the results) and 2) the tests were not conducted on two species of distinctly different stages of evolution for purposes of comparison.

Essentially what would be attempted if such tests were conducted today is to measure the WLD as being equivalent to thought level in order to open a quantitatively comparative window into the 5th dimension for the first time in human history. Where the rubber meets the road is whether the testing confirms or rejects the predictions. One of the lessons we can take from Dr. MacDougall as one of the hurdles to overcome in such testing will be keeping test subjects perfectly still so that movement does not interfere with registering accurate measurements. Possibly a paralytic substance can be administered prior to euthanasia of animals, and in regards to people I have no idea what may be acceptable or not..

Individual consciousness exits its host life-form upon death due to the equal complexity of thought levels between the mind and consciousness being broken by the quick drop towards zero in the life-forms ability to support its DNA based thought level. As the life-forms consciousness exits, there should be a measurable weight loss in grams or portion of a gram. At some later point in time, the consciousness that exited, in a manner not yet understood, connects with another life-form to continue its thought level ascension.

What is this idea of equal complexity? The idea is, that the body is providing a medium, an instrument for the consciousness (when, at some time before birth, it fuses with a life-form for its lifespan) utilizing it for ascension to higher thought levels. The only way that can occur, is if both are at the same level, i.e. in sync. The ability of the DNA to support a certain thought level must be at the same level of complexity as its companion consciousness thought level. In other words, there is a positive feedback to maintain equal physical and non-physical thought levels. If the DNA improves to increase the physical thought level, then the consciousness must increase to match it, and vice versa. Simple enough conceptually, yet it represents an amazing interaction.

Now someone might disagree with that notion, suggesting that a life-form early in development might not indicate clearly its potential DNA thought level it will later possess upon full maturity. However, it is asserted that equal complexity has to do with how the DNA matches up with the consciousness. As long as both are equal or near equal in complexity signature, they become self-aware, maturing in thought level in unison as the life-form matures. It is as if the consciousness is waking up again in each new life as thought levels increase.

Anyone can pontificate a theoretical dance of intricate thought provoking concepts, where no proof from predictions are provided. That is why the truly great theories always contain within them predictions that can be independently tested to verify the validity of the Theory behind them. With this in mind, a variety of potentially provable predictions have been established. They are the Consciousness Predictions:

One: *'The higher the WLD the higher the thought level, proportional to observed individual behavior.'*

Due to the symbiotic positive loop exchange between species evolution and individual consciousness evolution, both are balanced at the same thought level in each and every organism, including plant-life and micro-organisms. Therefore, the WLD (weight loss at death) for an individual life-form will be the same as its DNA supported-consciousness thought level (while it was alive and for that consciousness as it continues its upward ascension to ever higher thought levels), verifying this Theory's Definition of Evolution.

As stated earlier, WLD's for one species alone will not prove this prediction. Since WLD's will be numeric, ratios between different species can be calculated and compared to animal behavior. For example, a WLD for an opossum might turn out to be 1/5th that of a Raccoon, thus we would surmise the Raccoon is operating at 5 times the thought level of an opossum, and if their respective behavior supports a ratio of proportional behavior, then proof of prediction is provided. Other such comparisons will help to not only verify the prediction, but initiate the filling in of a Periodic Table of Consciousness Thought Levels, the parallel to the Periodic Table of Elements. In this manner the compression of both forms of matter via the 4th dimension to different thought levels and densities will be evident.

This idea of testing for WLD's from two different species at distinctly different stages of evolution; the Opossum and the Raccoon, I suggest test results could be announced on an international TV special, with two camps present and watching from around the world. Those that assert to an accidental universe or for some other reason being sure the test results will **not** confirm the first prediction listed below, and the other camp that is equally certain the prediction **will** be confirmed. It would be an exciting evening because as we all know this is a divisive subject with such testing results having far reaching implications regarding our understanding of the universe. I of course would be present as the cheerleader for those that expect the prediction to be confirmed and the other camp could be represented by Bill Nye, the Science Guy or some other scientist asserting to an accidental universe. Just prior to the announcement of the results the tension would be huge! And what would happen in the aftermath? Well, I guess we will just have to wait and find out which way the testing goes.

WLD as a measure of thought level is a stored form of energy (thought at rest), that can be inserted (in some numeric form) as the consciousness (co) quotient in the equation $t=coC^2$. Species WLD ranges will become part of a new field of Biology shown on a periodic table of consciousness thought levels. On the right hand margin will be color coded bars to designate brain layer thought level, i.e. R-Complex (reptilian), Limbic System (Mammalian) and Neo-cortex (dolphins, primates, particularly humans) for those sections of the chart that apply. Individual life-forms tested will determine a range of WLD readings for each species. WLD measurements will 'quantify' the average thought level of numerous different species, establishing a thought level hierarchy.

Two: *'Irrespective of physical size, the more evolved species will have a higher thought level/ WLD.'*

If we compare two species at different stages of Darwin's (physical) evolution, the smaller species that is more evolved, like humans (possessing the 3rd brain layer, the Neo-cortex) will have a greater WLD than a larger, less evolved species such as a horse (with the 2nd brain layer - Mammalian). E.g., a person weighing 150 lbs. will have a higher WLD than a horse weighing 1500 lbs. Sounds uncanny that a life-form weighing only 1/10th as much as the other would have a greater WLD, however this is the wrinkle in the Universe that will enable us to compare the thought levels of different species, establishing a consciousness weight scale, opening a window into the 5th dimension for the first time in human history.

Similarly, the WLD of a larger (in weight) opossum, will be much less than the WLD of a smaller (in weight) raccoon, due to a marsupial being less evolved than a mammal. A rhesus monkey will have a higher WLD than a Rhinoceros, a chimp's will be greater than a hippo, a dolphin's greater than a great white shark, a spider greater than a beetle and so on. What eye opening interesting comparisons it will be when WLD readings are coming in for different species, particularly for biologists and zoologists.

Three: *'Within each species, physical size of individuals will be a non-factor in determining thought level.'*

Consciousness compresses as it reaches higher thought levels, with physical size as a non-factor. For example, Mother Theresa was a very 'loving' person that weighed about a hundred pounds however her thought level was per this Theory much higher than the consciousness in most people, whose average size is much larger. Also, Einstein with his great 'intellect' must have had a very high thought level, but was not a physically sizable person.

Four: *'The two spectrums of consciousness ascension to higher thought levels are; knowledge gained, Intellect, and depth of connection with other consciousness, Love.'*

It is stipulated that the easiest way to confirm this prediction will be to attain WLD's from people, because the range of thought levels amongst people differs to a much greater extent than other species. Also, it will be much easier to document each person's life as it pertains to educational level, intellectual interests, special talents, and ability to form close connections with others. What should occur is the test results will fit into a hierarchy of WLD's that coincides with information about the individuals tested, regarding their intellect and depth of connection with others. Presumably these tests would take place in a hospice where permission is provided. The concern I have here is the nature of the event carries with it (rightfully so) such concern for respect to one's passing, that the very idea of conducting such tests may be rejected, holding back human consciousness from potentially moving forward. On the other hand, proof of a WLD, and more importantly confirmation of this prediction, will establish grounds for acceptance of the existence of consciousness, as some refer to it a person's soul. That confirmation would certainly provide great comfort to family members that were previously certain there is no such thing as a soul.

Intellect correlates to any and all information about the inner workings of the Universe, from the mundane to the lofty. As stated in the chapter, 'One Thought Leads to Another', what might seem like it would not be important, Love, is because without it consciousness could never come back together again to think as One to initiate the Big Bang. Tina Turner sang, "Love's nothing but a worn out emotion." Contrary to that lyric, we can easily observe closer social interactions between higher order animals, clearly indicating that higher thought levels carry the capacity for greater love. Love ranges from minor connections at the microbial level, through a spectrum of different possible love connections, to the full flowering of love for all consciousness in the Universe at the highest level. It is the God level (big G because all consciousness is in one) that provides the Love state providing for the unification of all consciousness to think as One, to release its thought at rest in the form of thermal energy into the mass to cause a Big Bang. As such, learning to love is integral to consciousness ascension and oscillation.

These two spectrums, Intellect and Love, apply to all life-forms however it will be easier to prove this prediction by comparing readings for people, due to our highest thought level when compared to other species. An experience during the development of this Theory led me to this prediction. In an effort to better understand how that arose, let's review what transpired:

As I started to contemplate the extremes of consciousness thought level, from the level in the primordial soup that initiated the symbiotic exchange between species and consciousness evolution, all the way up to the god level, I realized intellectually that all life-forms have the same motas, i.e. the same form of matter (consciousness) as we do. I happened to be driving at the time, and as I drove past a small farm house along a hilly, mountainous region, I could see in a fenced area behind it a small donkey following closely to the side of a large horse, and I was moved by their companionship. With this view came the thought of 'understanding' that they have the same motas of consciousness as all life forms do including people, and in that sense we are all the same, just at different thought levels. As this became 'intellectually' clear, and while the horse and donkey were still in view, a Falcon, with its hooked beak, flew directly across in front of the vehicle, turning to look straight at me. Our eyes made a direct connection, bringing with it a feeling of all four of us being strongly 'connected', convincing me without any doubt that we share the same form of matter, consciousness, and with it an associated feeling of deep connected 'Love' occurring, eliciting a sensation of passing through a plane to a higher, more energetic thought level.

It actually felt like I was jolted forward at a very high rate of speed, like accelerating to Starlight in a Star Wars movie, as all the pixels of my eyesight appeared as lines streaking, as I seemingly was being shot into the far distance. It caused me to hit the brakes to avoid getting into a wreck, with the tires screaming and the sensation immediately ceased. It was a funny moment to reflect back on because the cruise control was set to 62 mph, and therefore no physical acceleration had actually taken place, except of course in my mind/consciousness. I surmised, the sudden sense of acceleration felt like the speed of light, must be associated with the compression to higher thought levels, as seen in the equation, $t = coC2$, (thought equals consciousness times the speed of light squared). Not that all or even most transitions to higher thought levels will be associated with a similar sensation of light-speed acceleration, but apparently when there is a deep enough level of new understanding, it can occur.

Revisiting that memory, as you can imagine in that consciousness ascension moment, there was a realization that greater understanding increased the intellect (knowledge of how the Universe works) but also emoted an emotional response of Love (depth of connection with other consciousness), initiating an increase in thought level across both spectrums, compressing my consciousness to a higher, more energetic thought level. In other words, each and every individual consciousness is constantly under 4th dimensional compressive pressure (Conscilution), which is the same dimension applying pressure on our bodies. So that force is always there ready for you to come to some new understanding that allows this compressive force to increase your thought level.

Another way to view a higher thought level is to simply view it as a step closer to god consciousness, from whence we came. Possibly, because both spectrums were evenly accessed, the experience of going Starlight occurred. However, I am certain similar compressions can occur in either the intellectual or love spectrums separately, with no sense of acceleration occurring. In fact, without realizing it, simply by virtue of reading this book you will invariably initiate consciousness compressions by better understanding how the Universe works. But also at some point in wafting your way through a plethora of information, you may find yourself feeling strong emotions. These emotions are very natural, since our emotions are our gateway to the Love spectrum.

And what better way to receive stimuli than as a living, breathing organism, with a brain and a full array of emotions? The conclusion I reached, was there are two spectrums of consciousness compression. The first spectrum ranges from the lowest level of understanding in the lowest level of consciousness that first infused into the first appearing life-forms, initiating the positive loop exchange discussed earlier, to the full intellect of how the Universe works at the god level. The second spectrum is the connection felt with other consciousness, from the lowest level in a microbe, to the god level having Love (deepest level of connection) for all consciousness in the Universe. Although these two spectrums are interconnected, each individual has room for deviations, meaning, one person may be more loving (Mother Theresa) or have greater intellect (Einstein), yet both could be at the same thought level.

Think of the mental part as the link to Intellect, and the emotional responses as the gateway to Love. It appears, they are physical manifestations of what consciousness requires ascending to higher thought levels.

Men and women, tend at times to be at odds with one another due to differences in viewing various situations. However, if they can bridge their differences, are they not rewarded with love? And isn't love, 'the depth of connection with other consciousness', one half of the consciousness compression to higher thought levels? Could it be, that the feeling of being in love is a built in reward system for thinking and feeling together in a more complete manner than is possible separately? But that would mean the very act of loving produces higher thought levels and in turn more love. A sort of built in albedo (positive feedback) system. As both parties give up their own personal (stubbornly held) position on a given topic to see a greater balance of that situation by way of including their partner's viewpoint, therein lies a slightly higher consciousness perspective, and we feel love. That requires letting go of one's ego attachment to being personally correct, and substituting the ego for greater understanding. In a sense, a preview of what it would be like to think and feel like a god, which will represent the complete understanding of all perspectives and the deepest love connection.

'Is it any wonder why we all feel a need to have close personal

relationships?'

Five: *'The higher the thought level of a species the greater the spectrum of WLD measurements.'*

For example, Humans presumably having the highest thought level, should exhibit the widest range of readings. The reason for this is, higher thought levels equate to greater complexity, and greater complexity should translate to a wider range of readings. My Father and I once visited the San Diego Wildlife Zoo, and while watching Zebra and Wildebeest graze, two women were talking up a storm, and he said to me, "Humans really are the most interesting specie aren't they?" I agreed with him without hesitation, meaning it in a positive way, due to people's diverse interests and behavior.

Expectedly, there will be many species having partial or close overlap in their WLD ranges. Whether or not the highest thought level individual bottlenose dolphin, ape, orangutan, bonobo or chimp could breach the lowest thought level range of human WLD's, will remain an open question until independent testing renders the answer academic. Such testing is not advocated, i.e. unless a member of those species is in a medical condition requiring clinical euthanasia.

Six: *'All life-forms experience joy and pain equivalent to their thought level.'*

One of the age old debates is whether or not animals feel joy and pain. Some like to think they do not feel anything, inferring their existence is pure fodder for our disposal. However, since life-forms are the medium with which consciousness ascends to higher more energetic thought levels, then every thought level is inherently equivalent to the capability to experience joy and pain, physical and emotional.

Similarly, the ability to 'assimilate' experiences, positive and negative, is personal to each individual and will have its net effect on the progress of thought level ascension. Accordingly, the individuals that succeed best in life do not allow disappointments or even outright failures to impede their progress. To err is human, so if you can muster up the strength, get back up on your feet, dust yourself off and keep on trucking in the direction of higher thought levels.

Seven: *'Regardless of the cause, the severity of autism is proportional to the percentage of consciousness not fully connected to the brain, and this will be evident by much lower than average WLD's for severely autistic individuals.'*

The reason why is because the portion of consciousness outside the body will not register as part of the WLD measurement. 'Lower than average' is specified as compared to the other human WLD's.

The question is; 'What is the relationship between DNA to support a certain thought level, and the consciousness thought level that could cause a person to transition into an autistic state?'

Recall the prior information on the symbiotic positive loop exchange between the thought level of the mind and consciousness. Each must be equal in complexity to support the same thought level. If one rises, the other must match that new level, but if something occurs to cause an imbalance and they are no longer at the same thought level, there is a danger the mind will not operate at a sufficient level for the consciousness to remain fully engaged. At some threshold of in-balance, the consciousness partially dislodges and a new lower balance between consciousness and DNA thought level is established in the body, with the other portion of consciousness dislodged, i.e. in the 5th dimension. One fully connected consciousness, with some in and some outside the mind-body. Which of course means the severity of autism is proportional to the percentage of consciousness not fully connected to the brain.

Someone might take the position that when consciousness is outside the body it is in the 4th dimension, not the 5th, however it is a 5th dimensional form of matter that is either in the 3rd dimension when fused with a life-form or it is in the 5th dimension when outside the body. The 4th in this case provides a dimensional conduit for consciousness to fuse with life-forms to experience the 3rd dimension for the purpose of ascension to higher thought levels, which is presumably a much faster and more interesting process than attempting to do so in the 5th dimension only.

As it pertains to my brush with autism, one could postulate I had merely found an outlet for the energy built up in my mind, but it would be more accurate to state;

'The game of visualizing and manipulating random configurations at a high rate of speed, increased the complexity of the DNA to support a higher thought level that matched the consciousness thought level, establishing a new balance between the two.'

Probably though, getting too close to the Sun so to speak as the consciousness temporarily breached confinement within the mind to the exterior, caused me to thereafter be partially autistic from the standpoint that my consciousness could thereafter partially exit and return into the body, i.e. flexibly autistic. It took me many years to fully realize this ability, which became more apparent as the development of this Theory progressed. This is how questions about the Universe are tapped into. It is not like doing a Google search, but rather a meditative state that offers up some information, but never simply lays it all out. One must be engaged, inquisitive and persistent in pursuing answers to deep questions about whatever topic is pursued. It is as if information is available, yet we are expected to play a determined, passionate role in the discovery process, rather than simply taking dictation. In this manner the individual gets assists from the 5th, but also deserves credit for manifesting information into a reality for others to access for their potential benefit as well. This collaborative process also insures the individual has a working knowledge of the information.

There is a controversial therapy that forces autistic children to connect with their therapist visually, which in many cases seems to help reduce the severity of this malady, or even cure the individual. My Father, Graham, a mechanical engineer, told me in his early childhood he could not make eye contact with people. His parents had taken him to a doctor who in repeated sessions to overcome this affliction forced him to make eye contact. As an adult when addressing someone he would rapidly flutter his eyelids, hesitate for a moment then start talking. My Brother Gary was a genius, always getting straight A's, a great artist selling his artwork in Kauai and Honolulu, and prior to that working for a time as an industrial designer for Atari designing pinball machines. When we were growing up together he would transition into a second state. Playing together like kids do, then suddenly changing completely and tuning me and everything else out. His face would lose all joy, no longer wanting to make eye contact he would concentrate all thought in one visual direction. Whether or not he was also partially autistic is difficult to confirm, but my conjecture is all three of us did not fall far from the same tree. Each of us somehow averting the greater depths of autism, yet capable of making a connection into the 5th dimension regarding design.

Using eye contact as a method of treating autism is controversial, because it has no clear basis in medicine, however it has great premise with this Theory. As the individual is engaged to make eye contact or is physically touched, that forces more of the consciousness (separated from the body in the 5th dimension) back into the (3rd dimensional) mind. With repeated efforts, coupled with efforts to increase brain (DNA) complexity by way of education, some or all of the dislodged consciousness could permanently reconnect, resulting in reduced or possibly no autistic symptoms.

Unfortunately, people determine their behavior in all aspects of life via sensory feedback. As the child screams or acts out in some way to the forcing of eye contact or being touched, people (by way of negative feedback behavior response) discontinue their efforts. Obviously anything that forces consciousness back into the brain will cause discomfort to an autistic person, however no pain, means no gain. Contact must be repeated no matter how negative the reaction, but understand that over time the reactions will lessen as more consciousness re-seats in the 3rd dimension, the mind. Of course this means not trying to jolt the child out of autism in a brief period, but rather over time, i.e. weeks, months. However, efforts must be multiple and daily.

Symptoms vary greatly between autistic individuals. Some are oriented with left brained capabilities, while others are right brained, but they all have the same common malady of not having both hemispheres fully engaged. Do you think that could indicate a lack of consciousness in one hemisphere or the other, meaning partially dislodged? Of course that conclusion can only be justified if this prediction proves accurate.

Consciousness passes no judgment on the form DNA complexity takes. It simply and naturally responds to the equal complexity rule. In my case, the consciousness did not pass judgment or attempt to reject the newly acquired ability to randomly configure objects at a high rate of speed. In fact, one could say it welcomed it due to a new, higher thought balance being achieved. As long as the mind to support the consciousness is equal in complexity, the consciousness will remain fully connected to the mind. There is no suggestion here consciousness could ever fail to keep up with alterations in DNA to support a higher thought level, only the other way around, with the possibility of the mind failing to adequately support the thought level of the consciousness. This is suggested due to the fact that consciousness is predisposed for ascension, with the flexibility of its form of matter in the fifth dimension, compressing to higher thought levels as opportunities arise.

One extreme way to view the idea of consciousness dislodging from the brain, (due to an imbalance of thought levels between the mind and consciousness) is the complete dislodging that occurs upon the death of a life-form. As the life-form's bodily systems are shutting down, the complexity of the brain to support a thought level descends towards zero, the Equal Complexity rule is broken and the consciousness breaks free from the life-form completely. In the case of Autism, it is a malady whereby there remains complexity to support a partial connection, with a portion of consciousness still connected to the mind, protruding outside the body--one consciousness, just not where it all belongs, i.e. in the mind. A middle zone we would not wish on anyone, due to the greatly reduced capability for consciousness ascension that can take place in a thought state in limbo with no clear direction forward. Think of it like a piece of machinery that lacks grounding. Only by having a sufficient negative grounding, can there be an equal positive charge to operate the equipment the way it was designed, or in the case of a human, sufficient DNA complexity to support the entire consciousness.

Along these lines, has it ever occurred to anyone that as modern humans break free from the simplicity of earlier civilizations, consciousness is requiring greater DNA supported thought levels? Think about the magnitude of complex stimuli entering a young person's mind in the 21st century, with television, video games, fast paced movies, hand held devices, etc. vs. the only tech available as recently as the early part of the 20th century, the radio.

Autism is a malady that occurs in childhood, not adulthood. You do not hear of people in their thirties or forties suddenly becoming autistic. This suggests the formative years' when the mind is developing is the most problematic time period for autism to occur. Making sure children's minds are challenged enough to increase their DNA complexity to support their consciousness thought level is job number one, while also engaging them visually and by touch to keep them well grounded in the 3rd dimension. In this pursuit, watching television probably does not help. As a medium, it represents a tremendous amount of stimuli without much in the way of helping to increase the complexity of the mind, that is, unless it is specifically used as a learning tool. Remember, the consciousness is also watching TV, and may be increasing in thought level ahead of the child's DNA. Accordingly, and although I have not sought data to support this suggestion, I would venture an educated guess that autism rates in developed countries probably far exceed those in undeveloped countries, due to differences in stimuli input. People would be much better off having children build things and use learning toys. Anything to increase DNA complexity to support a higher thought level.

This does not infer that causes such as genetic predisposition cannot cause autism... as the prediction states, 'regardless of the cause'.

One possible cause of autism that has been at the center of great controversy is the use of mercury in immunity shots. Although recently discounted by the A.M.A. as a cause of autism, there is reason by way of this Theory to question that conclusion. If you are interested in the effects of mercury on the brain do a Google search, as there are too many negative effects to go into detail here. Suffice it to say, any injury inflicted on the brain, via for example mercury based inoculations, could in conjunction with other factors potentially act as a tipping point to reduce DNA thought level support, sufficient to cause a dislodging of a portion of the consciousness, which of course is autism. Why risk autism by use of mercury? In Norway it is now the law that mercury cannot be included in inoculations. They established this law because a study of theirs showed a 67% reduction in autism cases alone just by eliminating mercury in immunity shots.

Since an autistic person's consciousness is partially outside the individual, i.e. part of it is in the 5th dimension, means each particular expression of autism has some connection to the 5th. The 5th holds the spectrums of intellect and love, and as such each expression of autism probably has some connection with that storehouse of information to one degree or another. However, the more severe the autism, the less capable a person will be in conveying that information. Conversely it could be stated that the higher the function of an autistic person, the greater their ability to communicate this information such as Temple Grandin (bovine handling improvements).

Eight: *'Domesticated dogs & cats will have a greater WLD than their feral cousins, due to multiple generations of interaction with a higher thought level species (Humans).'*

The inference here, is by continuing its evolvement in the company of a higher thought level species, the domesticated dog or cat's thought level is positively influenced by the deeper connection with its master than the connection it would experience in the wild, in which strong survival pressures invariably limit the depth of connection with others of its own species. This prediction relates to the spectrum of 'Love', but might also positively influence 'Intellect'. For example, some dog breeds can learn the meaning of hundreds of words, and dogs trained to assist the police or used by the Defense Dept. are probably pushed to higher thought levels than normally expected in the wild.

In this scenario, a higher thought level does not necessarily mean greater survivability. In fact, it may mean the opposite, i.e. it only gains advantage if it translates after the omission of the species that helped it gain a higher thought level. Dogs and cats have been indoctrinated to follow our lead, but that only provides sustenance as long as Humans endure. Should something unforeseen occur to Humans, domesticated dogs and cats would probably be at a disadvantage to their feral cousins.

Nine: *'Due to the interconnection of the two spectrums of consciousness ascension, love cannot lag too far behind the ascent of intellect, thus contact with alien life-forms that are operating at a much higher thought level should not be the subject of great fear.'*

It is probably a more likely scenario in which they will show more caution in their contact with us, knowing our level of love is not as advanced as theirs.

Ten: *'All life, including laboratory created life forms, low in complexity like bacteria, or high in complexity like humans will be infused with consciousness at an in determinant pre-birth stage of cellular* *development.'*

In other words, consciousness does not pass judgment on the origin of life, instead simply fusing with it due to equal complexity. The exact moment this occurs during gestation is unknown and not predicted here.

Particularly interesting from this standpoint, is the idea of replicants (laboratory created humans) depicted in the famous sci-fi movie, 'Blade Runner'. There was tension between Harrison Ford and Sean Young's characters as to how a naturally born individual should interact with a replicant. At some point she saves his life by shooting another replicant, and presumably due to this incident he then sees her as more than just a scientific curiosity, gaining both respect and connection with her. Another more poignant moment of consideration for the comparison of 'natural vs. replicant' occurs when Roy Batty, the leader of the replicants is dying on a rainy rooftop, just after sparing Ford's life. In the original theatrical version (my favorite - not the Directors cut) Harrison Ford does a voice over in which he says;

"I don't know why he saved my life. Maybe in those last moments he loved life more than he ever had before. Not just his life...anybody's life...my life. All he'd wanted was the same answers the rest of us want.
Where did I come from?
Where am I going?
How long have I got?"

Based on this Theory the answer to those three biggest life questions, are:

Like cells in an orange, we 'come from' part of a God consciousness that transferred its thought at rest into the mass, to experience thought level ascension in successive lives in more evolved species.

We are rediscovering ourselves as we are 'going' back to that same level of consciousness.

We have 'got' tens of billions of years to either ascend to the god level, or as a default be drawn into a field of influence forcing ascension to that peak thought level.

What better way to acknowledge someone's existence than to attribute those biggest life questions to him? Later Ford and Young drive off into the sunset to live together as a couple (and supposedly the four year life limit for Nexus 6 is no longer a problem; don't you love the convenience of Hollywood?). So evidently he came to a place where the divisions of human vs. laboratory became so diffused (in part because of the idea of consciousness exists in all life), those differences no longer mattered, which was apparently one of the underlying messages of the movie.

As for reasons why the voice over by Harrison Ford is controversial must be considered by many, including myself, sociologically fascinating. Maybe in the sullen voiced manner in which it was delivered, it turned many off. However, in light of this prediction as it relates to the Theory, possibly a greater percentage of people will begin to side with the voice over theatrical version.

Along these lines, it would be very interesting from a perspective of behavior, to observe how a laboratory created human would conduct itself versus the clinical view of how we think it should act. Meaning, it would be much more interesting than probably expected. It would exhibit personality traits we usually hold unique for our species only. An individual consciousness having taken a similar journey as the rest of us will have fused with that body, and therefore not be the cold, disassociated, dispassionate type we might have expected from a laboratory created life-form, such as a replicant.

Eleven: *'All life-forms possess consciousness wherever they exist in the Universe.'*

Every life-form, big and small, wherever it exists in the Universe will possess consciousness. This is a fairly obvious prediction based on the implications suggested by this Theory, and indirectly from the Anthropic Principle.

Additionally, the conventional wisdom is that life is unique here on Earth and DNA evolved specific to this planet. However, the design of DNA is so inherently perfect for life-form expression it seems too great a reach to suggest that each and every planet evolves its own specific design. As I am sure you can agree, designs like DNA are not easily duplicated.

Once having arrived at what I felt was an explanation of existing Universe design, the realization that it is likely beyond our comprehension to know how consciousness came into existence, how it achieved the god level or in what manner and from what source it generated an equal amount of mass, came to mind. The real miracle is the coming into existence of either form of matter in the first place. What is the origin of something that seemingly came from nothingness? On that question, I am vexed, stumped.

That completes the consciousness predictions, but something that needs to be emphasized here, is verifying beyond any reasonable doubt this Theory and its predictions, will require numerous tests by more than one independent source.

'The first prediction is the keystone, the wedge shaped stone at the top-center, holding the archway of this Theory in place.'

Without verification of that prediction, all other aspects of this Theory remain on hold indefinitely. However, I have full confidence independent tests will confirm the first prediction correct. Given the importance of this prediction, it would not be right to simply accept one testing source's results. There is always the possibility an institution with a predisposed rejection of the Theory could orchestrate unfounded test results. Therefore, only multiple source testing will (in my opinion) confirm or deny the first prediction.

In particular, tests conducted for prediction number four will carry the most weight from a humanistic viewpoint, as it seeks to establish the two parameters of consciousness ascension, Intellect and Love, and with it the implication that each one of us is ascending in the direction of god consciousness. Independent tests taken by a scientific team at a hospice, upon the natural conclusion of individuals lives to determine the validity of this prediction, would certainly earn those members a Nobel peace prize.

Tests for different species of animals will contribute to the field of biology to assist with better understanding as to the relative thought levels of different species. Presuming humans are numero uno, then what species is second? Is it an orangutan, ape, chimpanzee, bottlenose dolphin or killer whale? And by virtue of determining their thought level, will we gain more respect for certain species by adding and enforcing more protection for them? Will we in part do this because of our realization by way of all species having consciousness, that they represent the graduated ladder to the Human level? And with that knowledge, our realization that without them would open a wide crevasse between lower organisms and the Human thought level, in turn asking for trouble as immature consciousness makes the leap from a very low thought level to a very high thought level in their first Human incarnation. With that in mind, do we greatly expand ecosystems specifically for the propagation of other primates? Do we establish a thought level cross over point, with any species exceeding that WLD number having full protection against human predation? For example, will we fully protect whales, dolphins, great apes and orangutans? Will we develop a greater appreciation for mid thought level species such as exotic cat species and seals?

How important do those forests and oceans seem now? In this regard, those individuals responsible for the making of the movie, 'Avatar', are acknowledged with accolades for their compassion, wisdom and effort in attempting to move consciousness forward to help more people understand the value of preserving our diverse ecosystems.

7. Atomic & Motaic Forces

Just as mass has atoms that can hold and release energy, consciousness must also have a form of matter that can hold and release energy, and the name given for this are motas. Mota is atom spelled backwards. Therefore, just as there are atomic forces there are motaic forces for the expansion and compression of consciousness in an oscillating universe.

The equation for mass compressing into a black hole is:

Gravitation (Gr) compressing a critical threshold amount of mass [M] (or mass coalescing into a black hole), underwhelming (U) the mass to minimal energy (e) descending its energy to cause the mass to become extremely cold, with temperature acting as the catalyst to cause the switch of atomic force roles from Strong dominant to Weak dominant (S~W), transitioning (=) the mass into a black hole (bh). The brackets around the letter M for mass, is the symbol used to denote the gravitational compression required to strip the mass of most of its energy. Einstein used a capitol E in his equation E=MC2 to represent (hydrogen) mass with peaked energy, and thus minimal energy in a black hole is expressed with a lower case e.

The nice part about having an equation for the forming of a black hole is it now makes it a simple task to write the equation for consciousness compressing into a god, which is:

$$Cn\ [co]\ O\ T\ S \sim W = g$$

Conscilution (Cn), the parallel force to Gravitation, compressing consciousness [co], overwhelming (O) to a peaked Thought (T) level, and as the Thought level peaks, it acts as a catalyst to cause the switching of the motaic force roles (S~W), with that consciousness transitioning into a god (g).

The switching of atomic roles occurs via energy loss or gain along with extreme temperatures, however, the motaic (consciousness) force role switching occurs via threshold thought levels, in which the energy for the peak thought level acts as the catalyst for the switch. The thought level in maximum compression is a peaked consciousness level and in the disbursed expansion is the minimal level, (a barely perceptible wavelength of thought energy). When force role switching includes the motaic forces, it is the atomic and motaic force role switching mechanism, abbreviated as the A&MFRSM.

Now you may be asking why a small g in the above equation to represent god? It's because, just as there are many black holes there are many gods. The implication is each individual consciousness has an innate capability to ascend to the god level. If more than one individual, e.g. Christ and Buddha can ascend to the god level, then we all can. Incidentally, the only time God exists with a capitol G, is when all consciousness has coalesced back into one God, predisposed for the next Big Bang. It probably amounts to less than one full second of time.

Just as a black hole has an 'event horizon', with any mass entering being coalesced into that black hole, then a god also has a similar area around it referred to in this Theory as an 'influence field'. Any consciousness entering a influence field will be drawn in, with its thought level will quickly transitioning through all the remaining thought levels (not previously achieved as an individual) to attain the god thought level. Just as mass can either compress into a black hole, or by default coalesce into a black hole, consciousness can either ascend to the god level or by default coalesce into that highest level, before or by the conclusion of the Big Crunch.

Now you may also be wondering why a black hole is at minimal energy, but conversely consciousness in a god is at peaked energy, yet both forms of matter are in their most compressed state. This is one of the amazing design aspects perceived while developing this Theory. As mass burns energy in stars, its energy is descending, with all mass energy in the Universe taken as a whole, moving in the direction of descent. Conversely, as consciousness ascends to higher thought levels via life-form experience, the overall energy of consciousness in the Universe is ascending, yet both forms of matter are also transitioning into their most compressed state to provide for an infinite flow of energy from Big Bang to Big Crunch, Universe after Universe. Thus, as overlapped single entities at the conclusion of the Big Crunch, one form of matter is at minimal (Black Hole) and the other (God) is at peaked energy. This is described as the;

'Differential energy expression of mass & consciousness'

A concept also expressed in the equation for the Big Bang (in the next chapter), a truly remarkable design parameter, requiring opposite forms of matter in opposite dimensions, having differential energy expression in their parallel compression and expansion. In other words, Gravitation and Consciculion are 'parallel forces', each compressing their respective form of matter, however each form of matter is energetically 'opposite' in response to being compressed, and thus are opposite in composition.

By proposing the 3rd & 5th are opposite dimensions, then raises the question of just how opposite are they, as in, to some degree or completely? The fact their respective forms of matter respond in a perfectly opposable manner in their compression, with mass releasing energy while consciousness gains higher more energetic thought levels (ectropy), indicates they are completely opposite. However, there may be another reason clearly pointing to these dimensions as being opposite, which has equal importance. The 4th dimension has a cohesiveness, a cross weaving, a lattice work of strong interconnection (as evidenced by the gravitational lensing of light), an intermediary dimension resulting from the expansion of two diametrically opposite dimensions.

Think about this for a moment; 'Your third dimensional brain and fifth dimensional consciousness are both in your head, balanced at the same thought level (via the symbiotic positive loop exchange), in opposite dimensions, with that dual connection made possible by way of the intermediary fourth dimension fabric of space-time.'

Differential energy expression establishes a flow of energy, with each form of matter consciously or unwittingly transferring its energy to the other. God consciously to the mass in the form of thermal energy to initiate the Big Bang expansion, and mass unwittingly to consciousness over the duration of the Universe, by way of sunlight for consciousness evolution via life-forms. I tried one day to figure some other way of oscillating energy, but quickly realized it was futile, because there is no other realistic or simpler way to achieve it.

However, by virtue of suggesting a way for energy to oscillate from one Universe expansion into the next, what has been put forth is an idea that on the surface may seem as though it is in conflict with the second law of Thermodynamics. That law states, "The entropy of the Universe always increases (meaning loss of energy), or equivalently, that perpetual motion machines are impossible." This law relates to mass, and yes, over the duration of the Universe mass is releasing energy. However, as this law relates to perpetual motion machines, it pertains specifically to the third dimension with one form of matter that can hold and release energy, mass. It does not state an equal amount of an opposite form of matter (consciousness), that can hold 'thought energy' in the fifth dimension reacting energetically in the opposite way to compression as mass, is subject to entropy.

'Although occurring independent of one another, as mass is relesing energy by way of compression to denser elements, consciousness is gaining energy by way of compression of consciousness (via Consclution) to higher, more energetic thought levels, with both forms of matter contributing to a differential energy expression, in which the energy of the Universe flows from one Universe expansion into the next, in the context of a perpetual motion machine.'

'As long as all consciousness concludes the Big Crunch in a peaked level God, the energy needed for the next Big Bang is locked and loaded.'

Keep in mind, when I say God, I do not mean some distant, mysterious entity, but rather I am referring to all of us. After all, we were all part of a God, with a capitol G before the Big Bang. I know that represents a leap of understanding, but this is the big picture being presented here. In fact, depending on the quality of one's life, there are many I am sure, upon a full understanding of having had their god level sacrificed to energize the mass and subsequently spewed out into space to start from scratch, might be resentful of the many hardships endured in this life and for the many they do not remember that occurred in previous incarnations (but nonetheless influenced their current life perspective).

I suppose existence during this sometimes troubling rise comes down to a Shakespearian philosophical question; "Is it better to have loved and lost, than to have never loved at all?" Fact is, we all love and take risks, yet often relationships and ventures do not work out the way we anticipated, but we always seem to find the courage to toss ourselves back into the fray to feel the rush of love and risk again. And when this Universe dance is over, we will feel so satisfied with our journey and accomplishment, we will collectively sacrifice it all again in a Big Bang moment of jubilation! Almost like a massive 4th of July fireworks display, but then the realization the journey must start anew registers, for we must start all over again from the beginning in a primordial soup of symbiotic evolution.

'Whoops!'

8.   The Big Bang Equation

For there to be a Big Crunch, all mass must either be part of a dying star that compresses into a black hole, or mass that gets drawn in and coalesces to become part of a black hole, and all black holes eventually combine into one Black Hole. Similarly, all consciousness either utilizes life-forms to ascend to the god level or consciousness is drawn in and coalesces as part of a god, with all gods eventually combining into one God. During the late stages of this event, the Big Crunch, it will become progressively more hostile for lower consciousness still infused in life-forms. As planets are drawn into black hole event horizons, lower consciousness will be extracted from those life-forms and coalesced into gods, acting as a default to accumulate all consciousness into one God by the end of the Big Crunch.

'The Big Crunch complete, is the final contraction and coalescing of minimal energy mass and peaked energy consciousness into two overlapping single entities, predisposed for the next Big Bang.'

On Christmas Eve 2005, I was just coming to the point of realizing mass descended in energy and consciousness ascended in energy, when there was a realization that once the Big Crunch was complete, both forms in their most compressed state, overlapped, would be predisposed for the next Big Bang. And then I realized that all that would need to happen was for God, which we will all be part of again, was to release its 'Thought at rest' in the form of thermal energy into the mass to cause another Big Bang. It was quite a moment of excitement, a wondrous Christmas gift.

For the Big Bang to occur, three sequential tasks must be undertaken. The first is to confirm and or adjust as needed precisely equal amounts of mass and consciousness, for the accurate expression of the 4th dimension once the expansion begins. The second is to calibrate the Black Hole to the correct spin rate so the Universe does not expand so fast stars never form or so slowly it collapses back into a God overlapping a Black Hole, and third, the release of a Thought at rest into the Black Hole in the form of thermal energy.

For the purposes of depicting an equation for the Big Bang, we will take it from the point when the thermal energy is released into the Black Hole. Let's first start by putting together the equation. The reason we are going to do this together is because it will be easier to understand if taken in small steps.

Let's start with the equation for the release of a Thought at rest. We will use the original equation t=coC2 (thought equals consciousness times the speed of light squared), but rearrange it to read,

GC2=T

God (G) times the speed of light squared (C2) equals a Thought (T).

The equation still works in the same manner, just as E=MC2 can be rearranged to mean the same as MC2=E. We now need to convert the 'Thought at rest' in the fifth dimension, which is released into a third dimensional form that will cause a chain reaction in the Black Hole to expand out into space. Since black holes are millions of degrees cold, to get the mass to initiate an expansion into its lightest, most energetic element, hydrogen, the Thought released is in the form of thermal energy. This is denoted as, r}te (meaning; released in the form of thermal energy), to instantaneously heat up the black hole mass to cause the switch in atomic force roles to effect the expansion. But we also need to transfer it into the Black Hole, so let's add >BH. Now our equation reads;

GC2=T}te>BH

God (G) times the speed of light squared (C2) equals a peak level Thought (T) released (r) in the form of (}) thermal energy (te), into (>) the Black Hole (BH).

At the point where the equation depicts a transfer of energy taking place into the Black Hole, it is natural to think in terms of something going from one place to another. But in reality, God is simply releasing its thought at rest in the form of thermal energy inside itself. However, since both single entities are overlapped, the transfer easily takes place into the mass. We might also think that must be a lot of heat for God to take inside itself, and we could ask how is that possible? The reason why is because God is in the 5th dimension, not the 3rd, thus no transfer of heat is conveyed except into the mass.

We will call that the short version of the equation to describe the Big Bang. However, it would be nice to have a more complete equation to indicate how both forms of substance, mass and consciousness, are changed by this sudden influx of thermal energy, to better understand how they expand out into newly generated space-time. So let's look at what happens to consciousness. We know from earlier in the Theory about the Energy Exchange rule that God consciousness or some part thereof is reduced to minimal thought, but we also know the Slater Tensor must be added to indicate the change to Strong force dominant. Thus, it would read;

UGtW~S

Meaning; underwhelming (U) God consciousness (G) to minimal thought energy (t), Weak motaic dominant (W) switching (~) to Strong motaic dominant (S).

Now let's look at mass. We have released the Thought at rest in the form of thermal energy into the Black Hole, so the mass has been heated up causing the expansion of atomic structure into hydrogen, the lightest, peaked energetic element, but we also need to add the Slater Tensor. So the equation reads;

OMEW~S

meaning; overwhelming (O) Black Hole mass (M) to peak energy (E), Weak atomic dominant (W) switching (~) to Strong atomic force dominant (S).

Now let's put the whole Big Bang equation together to read as follows:

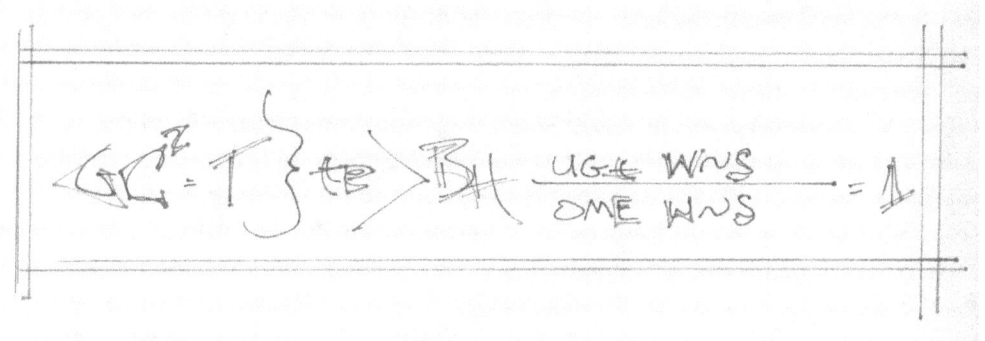

$$Gc^2 = T \} \, t\varepsilon \, \rangle \, BH \, \frac{UG + WNS}{OME \, WNS} = 1$$

God times the speed of light squared equals a Thought released in the form of thermal energy into the Black Hole, underwhelming God to minimal thought energy, Weak motaic dominant force switching to Strong dominant, (expanded state), (then below the line), overwhelming mass to peaked energy, with Weak atomic dominant switching to Strong dominant, (expanded state), equals 1 Universe (with 3 dimensions). It is made up of expanding 3rd, 4th & 5th dimensions, but they are intertwined as one Universe.

It is a perfectly good question to ask why the equation for the Big Bang is written with consciousness over mass. It is because when you write an equation with equal amounts above and below a line, they always divide into one another equaling 1. We could just as easily place mass over consciousness and the equation will still produce the same results. In this case, it means one Big Bang results in 1 expanding Universe. It makes no difference that mass is peaked with energy and consciousness is at minimal energy. What is important is that both forms of matter equally contribute to the Big Bang, sharing each expansion equally and interdependently. The inference is that since there are equal amounts of both forms of matter, it is a finite Universe.

Something that intrigues me is the constant multiplier to determine energy in mass (or thought power in consciousness) just happens to be the speed of light times itself. If this was an accidentally occurring Universe, why would the speed of light squared be so integral to determining the energy in either or both forms of matter? The only answer I can come up with is light must have 3rd & 5th dimensional properties, mixing as an amalgam in the 4th dimension. (For more on this phenomenon, see prediction number eight in The Consciousness Predictions chapter regarding the double slit tests.)

Now try to catch up to this mind warp, regarding mass only for a moment. There is no difference between mass expanding in the Big Bang and its contraction into a black hole, except they are opposite events, with one proviso, the contraction into a black hole represents less mass. How can it simply be an opposite event one might ask? Because the switching of the atomic force roles and the energy states are occurring in the opposite direction and order. This can be checked by comparing the equations for the compression of mass into a black hole and with the equation for the expansion of mass in the Big Bang. Again we are served well by the Nevell Tensors, which assist by providing this insight.

Here is another interesting notion. Mass expands into a new Universe for physical reasons, but how would consciousness in the 5th dimension expand out into space-time with mass?

I thought it might be interesting to explain how an answer to this question was achieved. It was actually one of the last missing pieces in this puzzle. I kept asking the autistic part of my mind to answer the question, but over a two week period it did not produce a suggestion.

I played the Big Bang expansion out in my thoughts, but kept getting stuck on how consciousness would move out into space with mass in the expansion. I then shifted to thought experiment mode, which is by way of imaged thoughts against a black background while lying down with my eyes closed. I cleared my imagination and conjured up a projection of God overlapping the Black Hole at the conclusion of the Big Crunch, asking; 'What causes them to overlap?' I figured an answer to that question might help with the other question. But there was no response. So I separated the two single entities, and asked what is it about them that make them different enough to draw them together and overlap? Some part of my mind placed a glowing + (plus) by God and a glowing – (minus) by the Black Hole. They actually looked like the plus and minus symbols were being actuated by a dimmer switch, as they lit up increasing in intensity to brightly glow, then dimming and fading away. I could not do that if I tried, so whether it came from the autistic part or somewhere else I have no idea. Previously in the development of this Theory I had used those two mathematical symbols to represent peaked energy (+) and minimal energy (-), the maximum and minimum amounts of energy each form of matter can hold, so it made sense that it would light up using those symbols.

I then thought, 'Ok, you gave me something. They are disparate in energy expression.' I then quickly realized it was no different than two polarized magnets, but in this case they are two polarized forms of overlapped matter. Thus, the Disparate Energy Attraction rule was born;

'The greater the disparity of energy between the two forms of matter, the greater their attraction.'

But more importantly, I also realized that in the action to cause the expansion, although their energy fields switch from God at peaked transitioning to minimal, and mass (Black Hole) transitioning from minimal to peaked, they remain disparate in their energy expression, and thus consciousness takes a ride with disbursing mass out into a new Universe. The answer ended up being quite simple, but as you can see getting to the answer took some effort.

9.   Expansion & Contraction

One of the big questions in Science is how mass expanded in the Big Bang, and again we are well served by the switching of atomic force roles. As thermal energy heats up the Black Hole, the atomic structure is overwhelmed with thermal energy, which expands the mass close to the breaking point, which acts as a catalyst to cause the switch from Weak force dominant to Strong force dominant, (expanded mass), and as each expanding atom pushes against all the other surrounding atoms, exerting a tremendous expansive force out into a New Universe.

The second expansive force, the centrifugal force of the spinning Black Hole, acts as an amplifier for the first expansive force. When stars such as neutron stars or black holes implode at the end of a stars life, they spin at a very high rate of speed (like a diver pulling his or her legs in to get more rotations). The smaller the space the mass is compressed into, the faster the spin rate. This secondary factor amplifies the speed mass and consciousness expands outward, effectively and sufficiently expanding the dimension of space-time into the Universe we know today.

Since according to physicists, the expansive force of the Big Bang needed to be precise to 1 followed by 120 zeros, so the Universe did not expand too fast for stars to form, or so slow the Universe fell back into a Black Hole, it is speculated here that once the Big Crunch is complete, God, (of which we were all part of and eventually will become part of again), precisely sets the spin rate of the Black Hole to insure a perfect (amplifying) expansive force, just prior to the release of the Thought release of thermal energy. And that presents an interesting question for a physics mathematician, which is:

'What was the spin rate of the Black Hole just prior to the Big Bang?'

What would the spin rate in rpm's have needed to be to cause the expansion in the manner that it occurred, taking into account all mass in one sphere, (Weak atomic force dominant) Black Hole, and the primary expansive action of atomic force role switching (to Strong force dominant) that occurred simultaneously? Certainly not an easy question to answer, but once answered, it will certainly provide an interesting insight into the expansive force required for the expansion.

Science's explanation for the Big Bang is it occurred from a spark from nothingness. Ok, then why would the Universe expand from a spark into a flat disc shape? It seems more likely that a spark from nothingness would simply expand in all directions. But it did not, because the Black Hole is spinning at the moment of the Big Bang and as it transitions from a solid orb, collapsing into a hot flattened disc spreading out into a new universe in a disc shape. This shape has been confirmed by Astronomical observation. Thus, the idea of the Universe expanding from a spinning Black Hole does have credence as a possibility.

Another big question we must ask ourselves is this; If black holes are eliminating space-time, then how does that affect the expansion of the Universe, and more specifically, would that change the shape from flat to some other shape? And if so, does that changing shape infer expansion forever into oblivion, or a Big Crunch?

Switching gears for a moment. If there are gods and black holes throughout the Universe, how do they overlap by the conclusion of the Big Crunch? For two reasons; the first is an idea we discussed earlier, which is they are both eliminating space-time, and the other is the, Disparate Energy Attraction rule. Since God is peaked with Thought energy and mass is at minimal energy at the conclusion of the Big Crunch, this causes an attraction and they perfectly overlap. This serves well for the transfer of energy about to take place in the Big Bang. Also, if at any time over the duration of the Universe, a god's field of influence enters a black hole's event horizon or vice versa, they will overlap.

Using that idea, here is an odd visualization. If a spacecraft entered an event horizon of a black hole and simultaneously a field of influence of a god at the center of a Galaxy, the mass of the ship would become part of that black hole and the consciousness of those on board would be transitioned to the god level. What a fast and furious trip that would be as your body is stretched until it physically dies, then your consciousness transitions through a myriad of ever higher thought levels into a peak level god. From excruciating pain and death to full knowledge of how the Universe works, with Love for all consciousness in just a few brief moments. What a rush!

10. Add One More Opposite Property

I have slipped the Opposite Sub-Verses comparison sheet in here again to add another category. The reason it was not included in Section 2 is because the information regarding energy oscillation had not been addressed. Now it is more appropriate to do so, we can better appreciate the significance of this addition listed first because all else follows.

Opposite Sub-Verses comparison

Initiated	Consciously	Naturally
Switch Tensors	W~S	S~W
Sub-Verses	S	W
Catalysts	Heat	Cold
Events	Inflation	Black Holes
4th Dimension	Generates	Eliminates
Electromgntsm	Ubiquitous	Non-existent
Gravitation	Distortion	Vacuum
Saves from	Annihilation	Singularity
Expression	Complexity	Uniformity
Existence	Secondary	Primary
Univ. Shape	Flat	Spherical

The decision to transfer a thought at rest in the form of thermal energy into the Black Hole by the God consciousness that is overlapping it is a **conscious** one, and is the only overt point in time in the oscillation of the universe in which we can clearly recognize the role consciousness plays as it *initiates* the Big Bang event. In the process there is a transition from the primary to the secondary sub-verse with all the properties that encompasses.

Should Opposite Sub-Verses come to pass as the prevailing Theory regarding universe design as it pertains to matter oscillation, and in the absence of test data for the first consciousness prediction, the conundrum for science will be how to find an accidental manner in which the thermal energy could have been delivered to initiate the Big Bang? The thought processes of determining if the Big Bang is a conscious choice or an accidental occurrence is compounded by the sheer magnitude of trying to come up with some explanation for how Opposite Sub-Verses, 'The Engine of Oscillation' could have accidentally originated. At a certain threshold of understanding for the framework by which the universe operates, it becomes extremely unlikely that it originated accidentally.

The switch S~W, of mass compressing into a black hole initiates at the long tail end of entropy, with near absolute cold **naturally initiating** the switch catalyst transitioning that mass back into the primary sub-verse.

Instead of fighting the idea of a non-accidental universe, I suggest science embrace this Theory sufficiently enough to warrant a thorough commitment to testing the first prediction of the consciousness predictions. That is, not reluctantly pursue a small test which will inevitably be labeled inconclusive, but rather take the testing procedures to their highest possible level of accuracy to insure conclusive results, positive or negative in relation to the prediction, knowing they may represent the best opportunity to verify the basis for energy oscillation.

11. Questions Answered & Provoked

As you have been privy, this simple idea of the atomic and motaic force role switching mechanism (A&MFRSM) or better understood as Opposite Sub-Verses has far reaching implications to answer many difficult questions. When you read each of them one after the other below, I think you will begin to see its importance, particularly in the context of oscillation and particularly regarding questions provoked by a singularity.

When considering which Theory most accurately depicts a given situation, (in this case the compression of mass into a black hole as a singularity vs. the atomic force role switching mechanism), the litmus test is often viewed as the one answering more questions than it provokes, becomes the prevailing Theory until a better one comes along. Accordingly, the following answers by A&MFRSM are presented:

 - Science has always wondered what caused the expansion of mass in the Big Bang. Switching from Weak force dom. to Strong dom., expands each atom against all the others to cause the expansion, with the spin rate (centrifugal force) of the Black Hole acting as the amplifier for the expansion, broadcasting mass out from its equator as the plasma flattens out into in a disc shaped Universe, a simple answer to a difficult question as to what caused the expansion?

 - In a sense, a role switching mechanism must exist to stop the mass from expanding to the annihilation break point in the Big Bang, and to stop its compression into a singularity. Metaphorically, the switch acts as the 3rd base coach so to speak yelling, "No, you're not going all the way. Stop here!" However, in this case you can think of the switching taking place just a few microns from home plate in relation to the full distance around the bases. Just short of annihilation and a singularity, yet sufficient to provide a minimum to maximum compression and expansion for mass. In the mirror dimension, the 5th, these switch points occur at minimal and peaked thought levels, yet the same premise exists, that these two forces stop the process short of causing damage to its respective form of matter, mass and consciousness.

- Ever since Einstein proved Gravitation was pushing down on celestial bodies by way of the distortion mass causes as it orbits in the fabric of space-time (the 4th dimension) via observation of gravitational lensing, Science has wondered what caused its creation, knowing presumably that it was somehow created out of the Big Bang expansion. The answer is the 4th dimension is an intermediary dimension created by an equal mix of the forces acting on both forms of matter, 3rd dimensional mass & 5th dimensional consciousness, expanding outward via the switch to Strong force dominant. Think of it as the closer on a door, applying a constant pressure to revert both forms of matter back into their state of origin, overlapped, maximum compression predisposed for the next Big Bang.

- In relation to Einstein's definition of Gravitation, black holes behave in odd ways which has created many questions. But, if by the switching of mass to Weak force dominant eliminates space-time in its mass (3rd dimension only) out to lesser 3rd dimensional extent until there is full space-time at their event horizons, then we come to understand that a vacuum gravitation exists, which is pulling in and eliminating the fabric of space-time causing a 4th dimensional void in that region of space, and many of those questions are answered.

- Physicists have always wondered why Gravitation is so weak. It is due to the switch of atomic force roles from the first sub-verse of origin (all mass in one Black Hole) to the 2nd, second sub-verse. Without a 2nd sub-verse relatively weak gravitational state, life-forms nor the intermediary dimension for consciousness to fuse with life-forms would be possible. Call this the king of all anthropic principles.

- The Universe is expanding in a flat disc shape at an ever faster pace, and many physicists view this to mean it will expand forever, rip apart and dim out. However, eliminating pockets of space-time (via the switch to weak force dominant) reduces the 4th's cohesiveness, no longer holding back energy from the Big Bang expansion to such a degree, causing the Universe expansion to accelerate. To have an oscillating Universe, there must be a mechanism to generate the 4th dimension and eliminate it, accomplished simply and easily via the A&MFRSM.

- Science has wondered ever since the discovery of the atomic Weak & Strong forces that exist in every atom in the Universe, why they possess these two forces which constitute 50% of the constant forces. The above simple answers to difficult enduring questions occurs from the switching of atomic (and motaic) force roles in the expansion and compression in an oscillating Universe. In so doing, the two states of Gravitation (& Conscilution) and their respective sub-verses provide answers to many questions about the universe.

- The extreme cold in black holes and extreme heat in the Big Bang are the tell-tale evidence indicating temperature (along with energy loss or gained) are the catalysts for the switching of the atomic force roles. The switch to Weak force dominant occurs 'naturally' via gravitational forces, whereas the switch to Strong force dominant is the only overt evidence of consciousness exerting its influence on Universe design over the duration of the Universe, as it initiates a transfer of its energy into 3rd dimensional mass, 'consciously'. Without consciousness to make that decision, the Black Hole would spin for eternity or until it finally slowed and stopped.

- Einstein, said some day when you will hear the right explanation for the design of the Universe it will be simple, because out of simplicity, arises complexity. Two switches is as 'simple' as it gets, but look at the flexibility in relation to oscillation that it supports with Opposite Sub-Verses. Established is 'complexity' in the expansion and the 'uniformity' in absolute compression, with two distinctly disparate states, each possessing unique but necessary qualities for the cycling of an oscillating Universe.

Questions provoked by a singularity:

- How did mass (and consciousness) expand into a new Universe out of zero volume? Did each vacuous atom expand against every other vacuous atom?

- How does an expanding singularity into a new Universe stop each atom from expanding to the breaking point?

- Due to the absoluteness of infinite density, would we not expect any transfer of energy into a singularity to be absorbed rather than altering it into expanding plasma?

- How does the parallel of a singularity which would be a zero volume God, contain the energy or the capability needed for the energy transfer in the Big Bang?

I would like to think some day it can be put to the test, either in an experiment or with an equation. If an experiment can compress mass (by removing most of its energy in an extremely cold, zero atmosphere environment) into a black hole, and there is some volume, then what happens when a tremendous amount of heat is applied to it by a laser? Does it expand into hydrogen? If so, then AFRSM is proven backwards and forwards, with the inference the switch also occurs in motas, via consciousness extremes based on the parallels between both forms of matter.

Short of conclusive proof, the mechanism of atomic force role switching is not proven by these questions answered, or by the questions provoked by a singularity. But, we can see that by answering so many difficult questions from a simple and possibly enduring idea, we must ask ourselves if the A&MFRSM elevates itself to become the current, prevailing Theory?

12. The Law of Proportionality

The Law of Proportionality:

'Each event/effect taking place in the cycling of an oscillating Universe must be proportional to the next, otherwise its forward momentum would slow to a stop'.

There are twelve events/effects for each oscillation. One does not start then the next one begin, but rather each effect has influence on the next proportionately like water pushing other water along a river.

'Is proportional to' is abbreviated as IsPto

1. IsPto release of a thought at rest in the form of a catalyst, thermal energy into the Black Hole

2. IsPto reduction in consciousness to minimal thought level and mass energy to peaked, hydrogen

3. IsPto switching of atomic & motaic force roles, W~S, S sub-verse only

4. IsPto an expanding, precisely equal combination of 3rd & 5th dimensional forms of matter and their paired forces, generating a mixed, intermediary 4th dimension

5. IsPto mass & consciousness distortion of the 4th dimension; soft push gravitation/conscilution compressing mass into a spectrum of less energetic elements & consciousness into a spectrum of higher, more energetic thought levels

6. IsPto switching of atomic/motaic force roles, S~W, via catalyst of near absolute zero degrees cold (black holes) and peak consciousness thought level (gods), W sub-verse (universe now composed of both S & W sub-verses)

7. IsPto increasing number and size of black holes

8. IsPto to reduction of 4th dimension cohesiveness

9. IsPto reduced resistance to inflation

10. IsPto accelerating universe expansion

11. IsPto changing universe shape from flat to a big crunch, spherical, W sub-verse only

12. IsPto overlapping, equal in quantity of matter, one at minimal energy; Black Hole & the other at peaked thought energy; God (composed of centilions of gods)

Rinse, repeat, cycle, oscillate…

Round like a circle within a spiral
Like a wheel within a wheel
Never ending or beginning on an ever spinning reel
Like a snowball down a mountain or a carnival balloon
Like a carousel that's running rings around the Moon…

From the song 'Windmills of your Mind' best sung by Noel Harrison

13.   Twelve Paradigm Shifts

Throughout our history, paradigm shifts of new understanding about our Universe have occurred usually with just one at a time with many years passing in between. However, now that we have opened the door the rest of the way to Universe design, we will collectively pass through a whopping twelve Paradigm Shifts, and they are:

I   Proof of consciousness confirming a 2nd form of matter.

II   Proof of the 5th dimension (from 1st paradigm shift)

III. Realizing all life-forms and gods have the same form of matter (consciousness) as we do, just at different thought/compression levels.

IV. Awareness of the symbiotic positive loop exchange evolutionary dynamic between life-forms and consciousness.

V. Transition from wondering if the Universe is the result of Randomness or Determinism, to knowing it is the latter.

VI.   Knowing how both forms of matter repeatedly expand and contract via the atomic & motaic force role switching mechanism.

VII   Go from wondering if the Universe expands until it rips apart, to knowing it oscillates.

VIII. Understanding we exist and transition by way of two sub-verses within one Universe.

IX. Knowing the Universe must harbor life on a scale never before fully understood.

X. The convergence of Science & Theology to more common ground.

XI. By understanding the differential energy expression of mass and consciousness, we transition from thinking the Universe will eventually dim out, to knowing it possesses an infinite energy flow.

XII.   Understanding all consciousness ascends to higher more energetic thought levels by way of Intellect & Love.

The oscillations continue to cycle from expansion to contraction, with the question we must ask ourselves;

'Will we ever get bored with this Universe, this playground we use for our consciousness ascension?'

But once the oscillations are in motion, obviously we cannot stop and change its design. We could only do that once the Big Crunch was complete and we are for that brief period of time thinking as one. Essentially the choices, without a lot of tinkering, comes down to two; status quo, i.e. expansion and contraction or overlapping single entities with no expansion. I guess the point is, there is not much else to do but to experience our own, as well as the others we are associated with ever changing circumstances in this immortal pursuit of ascension back to whence we came.

Triangle of Universe Design

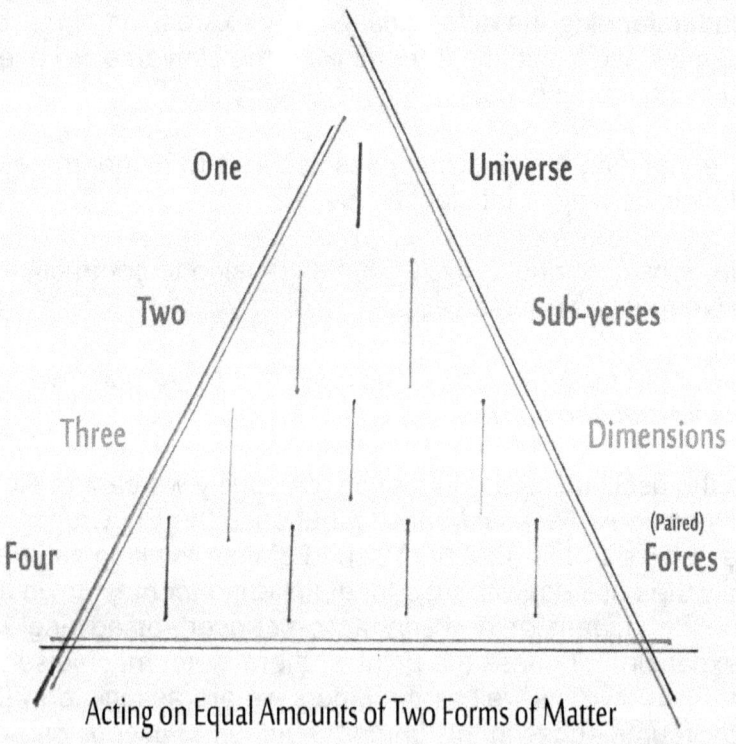

One   Universe

Two   Sub-verses

Three   Dimensions

Four   (Paired) Forces

Acting on Equal Amounts of Two Forms of Matter

14. Remarks

Scientists, theologians and others inclined to inquire, will come to realize by way of this Theory that we live in a finite, oscillating, Universe composed of opposite sub-verses, three dimensions, with equal amounts of two forms of matter that can hold and release energy, generating an intermediary dimension reflective of their parallel forces spread out over space and time, their forms of energy differentially expressed, their forms of matter attaining absolute compression and expansion via the atomic and motaic force role switching mechanism.

Finite: because there must be equally finite amounts of two forms of matter to generate a precisely even mix, intermediary dimension, the 4th, suitable for 5th dimensional consciousness to fuse with 3rd dimensional life-forms. Any Universe with unequal amounts of both forms of matter would generate an unequal mix point dimension in which reality would be skewed towards the portion having a greater amount of matter, rendering the fusing of consciousness with life-forms unlikely, causing an energy imbalance between the two realms, failing to achieve its intended purpose of a viable differential energy flow as evidenced by both forms of matter concluding the Big Crunch at different time intervals.

Oscillating: because once consciousness is confirmed, it is understood there must be a fundamental manner in which these two forms of matter expand and contract, et al, the force role switching mechanism, providing two sub-verses, because the sub-verse of expansion can only occur out of the sub-verse of absolute compression.

Segueing to your personal experience, did you find while mentally digesting the various aspects of this Theory it was demanding of your attention, such that you needed to take breaks between chapters, then come back to learn more later? Even though I wrote the Theory, I still cannot read it straight through, suspecting the reason is because in effect we are peering into a window not just of the 5th dimension, but on a bigger scale, Universe design itself. Based on the fourth consciousness prediction, this makes sense from the standpoint of assimilating knowledge gained, Intellect, as well as gaining insight into the value of deeper connections with other consciousness, Love. What is happening is our consciousness is compressing to higher, more energetic thought levels, but also in so doing, requiring DNA alterations to rebalance to its level. Those DNA alterations require a lot from the body and for most people those changes cannot occur in one read through, but rather in increments. Probably also though, just as I have discovered repeated readings continue to push my consciousness higher, you will most assuredly continue to get more consciousness compressions by reading the Theory at later points in time. Go back to it in a month or two and you will probably find it comes easier, but also there continues to be the need for breaks between chapters.

If you are an individual that has read this far, I tip my hat to you as someone with a bright, determined, inquisitive intellect. In a sense you have scaled a very high mountain. People say they climb high mountains because they are there, but in reality they do so because of the tremendous feeling of accomplishment in the face of a great challenge that is felt upon cresting the peak. The view from the top of the mountain into the valleys' below, along with its unique panoramic skyline, fills the spirit with a sense of all things possible, just as you are now feeling. You are amongst rare individuals capable of great accomplishments, with the willingness on your part to have delved deeply into Universe design.

That concludes both halves of the Theory of Opposite Sub-Verses. The book now transcends into many interesting and important philosophical perspectives in relation to this theory, that are integral to incorporating this information into your everyday life. It also includes challenges to all of us as a species as we move into the future. From pontifications on wisdom and creativity, to the convergence of science and theology, life beyond Earth, the challenge of ascent, paradigm shifts past & present and much more, it's time to bring it all together.

"May you enjoy many consciousness compressions", C. Nevell Slater

15. An Assortment of Pontifications

A Poem Depicting an Oscillating Universe, in 1791

Erasmus Darwin, an ancestor of Charles Darwin, wrote a book entitled, The Botanic Garden, with this poem.

Roll on, ye Stars! Exult in youthful prime,
Mark with bright curves the print-less steps of Time;
Near and more near your beamy cars approach,
And lessening orbs on lessening orbs encroach;
Flowers of the sky! Ye too to age must yield,
Frail as your silken sisters of the field!
Star after star from Heaven's high arch shall rush,
Sums sink on suns, and systems crush,
Headlong, extinct, to one dark center fall,
And Death and Night and Chaos mingle all!
Till o'er the wreck, emerging from the storm,
Immortal Nature lifts her changeful form,
Mounts from her funeral pyre on wings of flame,
And soars and shines another and the same.

The poem speaks for itself as an intuitive and excellent rendition of the New Universe through the Big Crunch and on to another Big Bang. One would think that such an idea of Universe oscillation could not have possibly been conceived of at such an early point in history. However, it is true that many an advanced and wise person down through the centuries has been able to understand this amazing cycling of the Universe on an intuitive level simply from deep meditation. Call this information part of the 5th dimension, or the collective unconscious with which we all can potentially draw from.

Densities of Matter, Parallel with Levels of Consciousness

Here is a 'fantastic' idea, in the true sense of the word, from information in the General Theory of Gravi-Conscilution, that tantalizes the imagination, yet on the surface seems too outrageous to be true and thus was not included in the body of the Theory. There is a parallel between the lowest form of consciousness and the lightest element, hydrogen. At the other end of the spectrum, there is a similar parallel between gods, the highest form of consciousness, and black holes the densest form of matter. On the periodic table, elements are ranked by their density, with hydrogen of course being listed first, but not listed are denser forms of matter, such as neutron stars and black holes due to a lack of their existence on Earth to more closely examine them. The point being, that since there is this apparent parallel between the extremities of mass densities to levels of consciousness, is it possible that every density of matter in the spectrum has a parallel consciousness level? Could there some day be a periodic table of consciousness? Could they be listed on one table? Thus, could Reptiles correspond to some element like tin and humans possibly to gold? Call it a wild idea that someday may prove to have a basis in truth.

It All Started With Creativity

Creativity is an integral aspect of any endeavor, in which one starts with a mere idea and progresses forward through an ongoing process of trial and error to hopefully eventuate with some form of success. One cannot create something new without the willingness to fail along the way, making numerous adjustments. What you'll find in life, is the most successful people are the ones that risk the most, but are the least affected by failures on the way to that success. This is the 99% perspiration which Thomas Edison talked about when it came to creating something new like the light bulb. He attempted over 1,000 trials to find out that a filament of compressed carbon inside a vacuum tube would illuminate with the right grounded voltage. It's easy to say and know the answer now, but to succeed in achieving a light bulb at the time took a willingness to repeatedly fail.

And it is this idea of creativity that intrigued me as a possible answer for the origin of the Universe, as part of this Cosmology Theory. We can read the Theory of Gravi-Conscultion and see the perfection in the way the Universe oscillates by way of two parallel realms with differentially expressed energy fields, providing an infinite flow of energy, and we could get real fanciful and think God created the Universe in one fell swoop. However, even the highest level of consciousness could not have known the exact workings of something as magnificent and complex as the Universe once it expands, having never created it before.

Thus, it was probably a case whereby a consciousness evolved to the highest level, God, and then got bored with that endless state and began experimenting in creative ways to develop a Universe that could expand and experience the various stages of consciousness evolution on a grand scale, then contract again. And thus a series of initial test Universes were probably created using mass/energy only, with no disbursed lowest form consciousness, to see what worked and what did not and then try, over and over again. How a consciousness, no matter how evolved might have created mass/energy, is an open question. In any case, this process might have taken hundreds of billions of years depending on the scale of the testing. When mass/energy and all of its physical components were suitable for the evolution of life, then tests were undertaken to test out consciousness infusing into life forms by disbursing part, but not all of its consciousness. Once it was understood exactly how to oscillate the Universe, from Big Bang to Big Crunch, over and over again with an infinite flow of energy, then the first full scale Universe was probably initiated in a Big Bang, and ever since it has been successfully oscillating.

At the end of each Universe cycle, as the gods combine and all remaining consciousness either evolves to or coalesces into an eventual single entity God, there is probably a period of calibration that takes place to confirm that both forms of matter are equal in amount, and the spin rate of the Black Hole is correctly calibrated prior to the Thought that takes place to initiate another Big Bang. Just like any system, the Universe is no different in as much as the head supervisor, our collective fully evolved consciousness, God, must make sure the system is set for another cycle before sacrificing the energy in the consciousness that created the Universe, with the ultimate faith that mass will compress to a point whereby consciousness can evolve into higher forms to sustain an ever repeating cycle.

Sacrifice

The notion of Sacrifice is a cornerstone idea in the Theory of Gravi-Consciution, as it relates to God sacrificing its highest form consciousness to infuse energy into the mass to initiate another New Universe, and in return the energized mass unwittingly sacrifices its peaked energy to support life forms for the purpose of individual consciousness evolution by way of burning that energy in stars. And, across the spectrum of human experience, sacrifice presents itself as such a potent word in our lives. People make sacrifices to support their children for education, on minor and major levels in their lives on a daily basis for a multitude of reasons. An extreme example of this was recently broadcast in a PBS special on mineral mines in the Bolivian mountains, in which the men that work the mines know they will die young, most between the ages of 35 and 40 from a disease called silicosis of the lungs, brought on by years of exposure to the silica in the air from pulverizing stone. And one of the men that knew he would die young, with humble, yet great pride, said they were sacrificing themselves for their families.

And over the centuries of humankinds' ascent, various religions have had thousands of examples of sacrifices to God for their well-being. I use to view these as purely pagan acts, driven from our culture's lack of education and understanding. Yet, through the development of this Theory, after realizing that sacrifice is such an integral part of our oscillating Universe, and seeing it now in many aspects of people's lives, it should give us all a more enlightened understanding of its significance. One might wonder if religious sacrifices are based on an intuitive understanding that God sacrificed its consciousness for us? Of course what has been lost in some of this ritualization, is the fact we were all part of the God consciousness that made that sacrifice. Nonetheless, it is noteworthy in respect to possibly altering our viewpoint more positively in regards to the rituals of sacrifice.

The Ultimate Act of Faith

Another interesting aspect of all religions is the idea of Faith. That even though we cannot know God with our basic senses, we must have faith to believe, which is called an act of faith. And this is a fascinating parallel to the faith that must be exercised when God sacrifices its energy to the mass to initiate another Big Bang. Knowing that its consciousness will be reduced to the lowest form of consciousness, with barely perceptible wavelengths of conscious thought, having faith that the Universe is correctly calibrated to support evolving life forms to allow for the process of consciousness evolution to initiate on planets like ours, and that this process will lead to an eventuation of one God again, is an act of faith at the most extreme level, ipso facto.

On a minor level of faith, imagine if a scientist said to you, "In this experiment, your consciousness will be transferred into a monitor lizard for about ten minutes, and its consciousness into your body, but have faith, because I will make sure it doesn't damage your body and that you will pop back into your own body soon enough. Trust me - have faith." Obviously by virtue of this extreme example, it is one of those words that takes on more significance depending on the level of faith required. But in this case, even though this test of faith doesn't even come close to the act of faith God must possess in the Big Bang, no one in their right mind would ever consider attempting a stunt that might leave that person for the remainder of that lifetime in a reptilian state, thereafter continually tasting the air with their tongue.

Contrasting this idea that all God consciousness was reduced to the lowest form, is the idea that the core of God probably did not expand into the Universe. The reason why is because it is too much of a leap of faith to risk the end of the Universe should anything go wrong in the development of the Universe to produce the god/God level again. Thus, the God core probably still remains in the original location of the Big Bang to insure as a default to coalesce all lower consciousness back into one God. If so, then an equal amount of black hole would have remained behind as well. Maybe astronomical observation can make that determination some day.

However this idea of faith is integral to religion, to the evolution of the Universe, and plays a role in every person's life, irrespective of their particular belief system. In fact, I have had firsthand experience with people who have no faith in the Universe or God or whatever they think might be out there, and have observed the ebb and flow of their panic, as each new challenge is met with overwrought dread. And thus, faith has its place in our lives. We must often have a sense of faith that on this journey all will go well. That something may or may not go right, but faith will be exercised, that for example, the aeroplanes involved in my transport will take off and land safely, the food will not be tainted and I will not get robbed or shot. Without any faith whatsoever, people would be frozen with fear.

Thus, we can see in the highest level of consciousness, God, the highest levels of creativity, sacrifice and faith, with those same parallels present in our lower individual thought level day to day lives. And by understanding these higher ideals, we can better understand the basis for theology, as it helps to convey those ideals to their following. It is as if religion on an intuitive level understood these aspects, and even though they are sometimes masked by contrived ritual or fables, we can see the truth of their parallel in the way the Universe presents itself.

Sustainability Level/ a Forward Look

Just as the compression spectrums of Intellect and Love are factors governing our thought level development, there is a parallel to this idea to humankind's population total. If we are indifferent to the negative forces we exert on our environment, that attitude will exhibit itself as a limiting factor on our population expansion. And by contrast, to the extent we can become an integral part of a healthy environment, will determine our maximum population possible. By healthy, meaning, biodiversity thrives, the atmosphere is stable i.e. not heading towards catastrophic global warming, plenty of fresh water supplies, ample energy, food sources and minimal disease. The challenge will be our ability to gear technology to aid us in getting what we need, but in a way that is minimally destructive to the environment. That is a tricky balance to achieve, requiring us to continue to be vigilant about our impact, to minimize its negative effects as we spread outward to fill the globe.

So far humankind has been on a quick track to achieve as much as possible with forms of technology that have been in many areas destructive to the environment. However, over time we have improved technology to be less destructive to one degree or another, but obviously greater strides must be made. It seems apparent we are now in a time period whereby we are realizing and accepting our impact on the planet and making adjustments accordingly. For example the Montreal Protocol is a worldwide accord to eliminate the use of CFC's to reduce the hole in the ozone layer. Since its signing, the ozone layer is already showing signs of stabilizing and in time will return to its former non-hole state. A great indication we can avert humankind made disasters if we are willing to take action collectively. The first step is to establish the science of whatever negative effects are occurring, then upon reasonable proof, accept information and take unified action to effect change. In this regard, the Montreal Protocol was a very promising step forward, because it represents hope that unified action on an even bigger topic can also take place to reduce greenhouse gases and avert what could become a worsening upward spiral of global warming.

In the case of global warming, it's not really a case of whether the science is right or not, because it is really just simple atmospheric chemistry. We have changed the chemical composition of the atmosphere, and as such more energy is getting trapped (greenhouse effect) in the atmosphere, transferring into the oceans and thus the global weather system must exert more force to disburse and balance out that increase in energy input. That means more dramatic weather events known as climate change. The problem is not the science, which is sound, but as compared to CFC's, the answer to this problem is the weaning of our economy from one driven with the use of fossil fuels (greenhouse gases), to renewable energy sources. The resistance is based more on a fear of what might happen to the economy as this transition occurs because growth is required to keep lending money, rather than any logical argument against something as easily understood as global warming.

You can think of humankind's technological ascent as being comparative to a bell curve. As it initiated, our technology was at the bottom of the bell curve, the most destructive to the environment, and as technology has improved we have moved up the bell curve. Right now we are at the peak of the curve. We are still using fossil fuels at an alarming rate for the majority of our energy needs, yet we are on the cusp of making the necessary transition to better technology that will start us down the other side of the bell curve. At the bottom right side of the bell curve we will be using more energy than at any previous time in our history, yet the production of the energy sources will have minimal negative effect on the environment. And that of course will not only mean a healthier planet for us, but also for all the other species as well.

One Lifetime versus the Duration of the Universe

Here's an interesting observation regarding the length of one human life versus the duration of the Universe so far, and for the remaining duration of this Universe cycle. If an average person lives say 75 years, with the Universe since the Big Bang being 13.7 billion years old, and is predicted to last between 30 -100 billion years, then the chances of being consciously aware of one's own existence in any one year of the Universe just as it has existed so far, is 75 in 13,700,000,000 or 1 in 182,666,666,000. If you compare those 75 years to the entire duration of the Universe, estimated at say 60 billion years, then the chances of you being consciously aware of your existence are only 1 in 800,000,000, (one in eight hundred million).

If you narrow the field to the period of mammals (the last 70 million years) then your chances are 1 in 933,333. Even better are your odds since Homo sapiens started evolving 4.5 million years ago, which is 1 in 60,000.

So the best chances of you being consciously aware of your existence after the Big Bang, are by waiting 13 billion 695.5 million years for homo-sapiens to start evolving, then you still have to pick the right seat in a stadium that holds 60,000 people at any one moment in that time period, otherwise you do not exist. That is of course if the Universe has only one dimension, the 3rd.

However, in the case of the Universe being designed as stipulated in the General Theory of Gravi-Consciction, then your chances of being consciously aware of your existence at any one moment of time, throughout the entire duration of the Universe is closer to 100%, absent time spent in a coma. It doesn't matter if you are infused into a life form or in between lives; you have an immortal consciousness that is slowly compressing in the direction of god consciousness.

This time perspective does not prove this Theory as being correct, however with all that time that has taken place and will continue to take place over the duration of the Universe, wouldn't it be much more preferable to be actively pursuing your individual evolution, versus just one fleeting incarnation?

Preciseness of Big Bang Expansion

If the Universe had expanded a little faster, matter would have sped into infinity and never formed one star. If it had expanded too slowly it would have collapsed back into one Black Hole and never formed a star. To strike the perfect balance between too fast and too slow, the force, something physicists call "the Dark Energy Term" had to be accurate to one part in ten with 120 zeros following it.

Obviously that is a staggeringly demanding requirement, especially when you consider each one of us represents a bit of consciousness that was part of the pre-Big Bang God, exacting that level of preciseness in Thought to initiate the Big Bang.

The magnitude of that Thought illustrates the multitude of greater consciousness levels we have to look forward to as we individually and collectively compress to higher thought levels. In a sense we have already done the hard part, i.e. waiting for the early Universe to develop galaxies and life forms, ascending via a multitude of life form perspectives eventuating at this juncture in human form, in a modern civilized time period. The rest should be gravy as we incarnate in genetically more evolved humanoids or whatever species avails itself for each of us to continue in this endeavor, in presumably ever more evolved civilizations.

Wisdom

One of the ideas wisdom teaches us is when two people are passionately arguing their divergent viewpoints, there is usually an element of truth to both sides of the argument, yet both are absent the truth of the other's argument. And we can see this idea reflected with the ongoing passionate battle Science and Theology regarding evolution and on a broader scale the entire Universe. Religion has up until now had the unenviable position of trying to make an argument with no verifiable proof, with Science understandably maintaining a solid position with provable experiments to back up their positions in all areas regarding the Earth and the Universe, that was up until the discovery of the Big Bang.

The Big Bang event itself has presented a problem for Science since its proof thereof having been provided by a number of scientists, due to the implications that some outside force must have caused this occurrence, because none of the four constant forces can account for that event. The discovery and proof of the Big Bang was the first event to have taken place in the history of this Universe, as far as we know, and its discovery by Science has given Theology a foothold with which to argue strongly for the existence of God.

With potential proof of the Theory of Gravi-Conscilution, by way of verifications from its predictions, a scientific explanation will provide an understanding that God isn't a separate entity from us, but rather we are all composed of the same motas of consciousness, just less compressed. We just happen to be the part of the expansion that was reduced to the lowest thought level for the purpose of being challenged and sometimes enjoying the process of evolving into higher forms back in the direction to that highest level, to once again take part in another Big Bang as one God.

So, out of this convergence to more common ground between Science and Theology, the latter will benefit from proof thereof, yet on the other hand the idea of Theology in a way will be reduced in some senses of the word as it becomes part of Science. Some of the mystery that has been the basis for so much disagreement over the years, and so many divergent theological viewpoints, will be replaced by a more generic understanding of the overall process of our amazing Universe. And with this transition, some of the need to have faith will have dissolved as Science will have put its stamp on facts with which we can rely upon, reducing the need to make a leap of faith.

And in this sense, wisdom will have won the day. In the final analysis, both sides of the argument will have held positions we can point to as being correct. As Science and Theology converge to more common ground, the differences between our respective views of God and the Universe will become less divergent, suggesting a gradual transition over time, as this new paradigm permeates the populace towards a more peaceful coexistence.

16.   Design Convergence of Science and Theology

At a certain stage of developing this Theory, I sought a naturally occurring design with which to compare to the master universe design models proposed by Science, Theology and Gravi-Conscilution. The most complex naturally occurring design I could find was DNA, with its two parallel poly-nucleotides strands with four repeating single nucleotides. The basic design of DNA is therefore:

2 primary with 4 (repeating single) subordinates.

At present Science only recognizes 1 form of matter that will hold energy, mass, with its 4 subordinate constant forces, Gravitation, Electromagnetism, Strong Atomic and Weak Atomic forces. This master universe design model by Science is therefore:

1 primary with 4 (constant force) subordinates.

However, if one were to presume the truth of a Universe supported by that design, then we would have to presume DNA's design supports a more complex system than the Universe. But that would mean a naturally occurring design arising out of its master design, evolved to become more complex. I suppose it is possible, but does that seem counter intuitive? Perhaps, we should define our basic position, which is;

'Any naturally occurring design arising out of a master design inherently will be less complex.'

But that would suggest Science's master design was incomplete. At the same time though we could presume that to be a true statement, since Science has so far been unable to explain what caused the Big Bang, or where the energy came from that infused into the mass.

Let's segue to the design suggested by Theology, which suggests everything is God, a form of substance we are referring in this Theory as consciousness. Now, without Theology realizing it per se', the constant forces under God are Consciulation, Conscimagnetism, Strong motaic and Weak motaic forces. This master universe design model by Theology is therefore,

1 primary with 4 (constant force) subordinates.

If we compare Science and Theology's master designs, we notice right away they suggest the same design, yet both represent a different form of matter. Science has mass, while theology has consciousness, yet both master designs are still less complex than DNA, which does not fit with our definition regarding naturally occurring designs.

Now let's consider Gravi-Consciulation's master design, which is two forms of matter (2 primary), mass & consciousness, governed by four pairs of constant force subordinates, Gravitation/Conscilution, Electromagnetism/Conscimagnetism, Strong atomic/motaic, Weak atomic/motaic forces. This has a basic design of,

2 primary with 4 pairs of (constant force) subordinates.

If we compare this design with DNA, we notice they both have two primary, therefore in that regard they are equal. However, four pairs of subordinates is more complex than four repeating single subordinates, which means DNA is less complex than the Master Design it arose out of, and we have satisfied our definition of naturally occurring designs.

On the surface, it might seem we had just intellectually converged Science and Theology. However, one of the recurring conundrums associated with an exercise like this, is the basic differing viewpoints of Science and Theology. For Science, the Universe occurred by way of an idea known as 'Randomness', the idea that things happen accidentally with no purpose. Based on Randomness, the Universe sprang forth from an accidental spark, and as such a naturally occurring design like DNA is just as accidental, and therefore could just as easily be more or less complex than the accidental Universe from which it evolved. Theology on the other hand bases their belief on a concept known as 'Determinism', which means there is a consciously intended design and purpose to the Universe by a God.

Randomness, is why Brandon Carter stated his Anthropic Principle (at Cambridge University, UK in 1973) by prefacing it using the word 'accidental' to describe the physics parameters providing for a Universe that 'just happens' to supports life, because that terminology would allow him to make that kind of suggestion in the field of Science. For those not familiar with English culture, which I am of course familiar with having been born in London to two English parents, Carter may have used language to convey a subtle form of sarcasm. In other words, how could a multitude of parameters just happen to provide for a Universe supportive of life-forms, without any intended purpose? It is kind of like saying, 'I was walking down the street when the sidewalk opened up, a box popped high enough for me to grab, and inside was a fortune in gems.' How beautifully and accidentally, coincidental!

But then again we should to take our collective hats off to Science for sticking with the need for everything it stands by to have been proven beyond a shadow of a doubt. In this respect, there is great assurance that what Science does put a stamp of approval on, is known facts with which we can rely. Beyond pure facts that have been scientifically proven, Science seems strongly geared towards theories that support a purely physical, accidental Universe, making it difficult to gain their audience to entertain other possibilities, like this Theory. I would like to think independent confirmation of prediction number one and four would permanently alter that position.

17. Reaching a Pinnacle of Technological Development

As we reach a pinnacle of technological development, along with a more accurate and complete understanding of our Universe, the capability to terra-form a nearby planet is within the same level of knowledge as reaching the point of being able to manipulate genetic expression, along with the search for life on extra solar planets using land based and space telescopes, achieving a net energy return fusion reactor, or the understanding of how the Universe oscillates. One is simply a reflection of having achieved the others, as we approach a stage of advanced development leading to a paradigm shift of understanding for our responsibility of consciousness evolution in this region of space.

We can see from this latest progression, humankind has come full circle, from that critical moment when one of our ancestors first stood up as a prey specie, to separating ourselves from the animal kingdom by becoming feared predators able to control fire, to controlling agriculture and farm animals, passing many major paradigm shifts of understanding of our Universe, to eventuate at a point of soon beginning passage through the greatest of all paradigm shifts.

However, if we substantially reduce biodiversity, we damage the graduated links that form the ascension of consciousness to the human level. If we cause great destruction to our own species, we set back the evolutionary process on our planet by thousands or possibly even millions of years. As we reach this pinnacle of technological advancement in which we could possibly insure the health of our planet and advance to occupy Mars, we are at the same time causing the degradation of the very planet and its biodiversity that gave rise to our thought level and our species over millions of years, and in a greater sense since Evolution first got started on Earth.

We are at a crossroads, staring at some very difficult decisions that will either move us forward in a positive or negative direction. However, knowledge is a powerful tool and there is hope that by understanding our Universe more accurately, clarity in the decision making process by those most able to affect change, will ultimately move this natural biosphere's destiny in a positive direction.

18.   Life Beyond Earth

The realization that a parallel amount of consciousness as mass disbursed outwardly from the origins of a God and Black Hole, is a direct indicator of just how much consciousness must be evolving in other parts of the Universe. Therefore, regardless of our failure so far to detect extra solar life, it is now clear from this idea of two equal forms of matter that:

'The Universe must harbor life on a scale never before fully understood.'

And here we have a corollary to the earlier assertion of a collective unconscious at work. Right now there are numerous projects underway using land and space telescopes, attempting to detect life on other planets, and at the same time this theory is being published with depictive equations to help clarify that the search for extra solar life should succeed. Both efforts are independent of one another, yet both are connected, as if a collective unconscious effort is taking place in this time period to find life beyond our planet.

If one were to argue against the idea of an equal amount of consciousness to mass in the Universe, by virtue of a lack of observable life from Earth, there are five possible counter-arguments that can be asserted, and they are:

First, we still have not reached the astronomical viewing ability to view many Earth positioned planets, i.e. those in the goldilocks range from its star, that perfect distance to support life. As such, we should not draw any conclusions yet regarding the possibility of life on other planets. The only planet in our solar system other than Earth in a suitable range from the Sun, having a composition that could support life with some terra-forming, is Mars, and it is still being analyzed as to whether it ever supported life or harbors microbial life.

Secondly, the Universe is still young, and therefore planets are still evolving in the direction of becoming capable at some future point of supporting life-forms on a greater scale than at present.

The third possibility is the idea that the consciousness evolution process reaches a crescendo effect, whereby the life forms on certain planets reach the capability of terra-forming suitable nearby planets for expansion of life-forms. This possibility has been discussed regarding Mars. So far, the one deterrent to terra-forming Mars has been the scientific pursuit of determining if life ever existed on Mars before compromising its pristine environment with Earth originated genetic based life-forms. Once the research is complete however, plans are to terra-form its atmosphere by way of plant growth in enclosures capable of living in a CO_2 rich atmosphere, which is estimated to take about 300 thousand years to reach the point of supporting life as we know it on Earth, and protecting its life forms from radiation from the Sun with the natural development of an ozone layer. However, technology is always changing, and in the future there may be ways to speed up the process. Keep in mind however, 300,000 years over the duration of the Universe represents an extremely short period of time for life to progress to a complex level of development.

In any case, if all planets having life-forms terra-form nearby planets, then individual consciousness has a much greater opportunity to evolve.

The fourth possibility is that the Conscilution process does not need a Universe chalk full of live planets for consciousness to ascend to the god level. Keep in mind that humankind only came out of caves a few thousand years ago, and therefore there is probably some future point in time when some life-form, human or a genetic improvement of humans, or maybe another life form that evolves out of a mass extinction, thinks at such a high cognitive level, that consciousness infusing into these life forms will ascend to the god level in just a few lifetimes. In so doing, these planets will act as a channel to speed up this process to a point where consciousness progresses to the highest level at a very fast rate. The question at hand might be, how physically evolved can a human become with genetic engineering in the direction of a hybrid species that operates at a much higher thought level? After all, with a greater understanding of the implications of greater physical evolution, having its symbiotic parallel effect in the consciousness evolution process, will we simply wait millions of years to find out if we are physically evolving to a higher level, or will we speed up the process with genetic alterations? This represents a critically important moral and ethical question for future debate.

A fifth more obvious possibility regarding mass and consciousness is that in the 3rd dimension most mass coalesces into black holes, versus being part of a critical mass supernovae star that compresses into a black hole. Therefore, most consciousness probably coalesces into the god level versus enlightening into that level.

The process is so gradual over such a long period of time, that on the surface it appeared initially to scientists, even as recently as the early part of the 20th century, to be a static Universe, meaning one that had always been here. With proof of the Big Bang, and now with this theory modeling a perfectly elastic and oscillating Universe, coupled with the understanding from past paradigm shifts that humankind has historically erred on the side of the least possibility, we must wonder when Science will have the proof needed to catch up to the conclusion that we live in an oscillating Universe.

19. The Challenge of Ascent

If you accept the idea from this Theory, that all life forms have a remnant bit of spirit from the original all-knowing Consciousness that caused the Big Bang, then in a sense isn't all consciousness in all life forms the sons and daughters of God? Motas on loan until we evolve, or are coalesce into that level. But that would invite the idea that all of us have the same potential as Christ or Buddha to ascend into an enlightened level, which is exactly what is being asserted here.

However, the process of evolving is challenging, no doubt. But what if it wasn't challenging?

There was a Twilight Zone episode, 'A Nice Place to Visit', with Sebastion Cabot, that catches the essence of that question. The story focuses on 'Rocky', a lifetime criminal always running from the law. One night his luck runs out on a heist gone wrong and he gets shot and killed by the police. Suddenly he's some place that seems really wonderful. He can have anything he wants. No matter what, he always knocks in all the billiard balls, wins every casino bet and always gets the girls he wants. He is living in a place designed just for him, but everyone except himself and his guide (Cabot) are simply conjured up as props to fill roles. At one point he sends his girls into a room to wait while he leaves with his guide, but then goes back to tell them something and they are not there. He is puzzled by their absence and starts to question his situation. He realizes to feel a little more alive he needs to lose at something, like a bet or get caught robbing a bank. But this entails his guide planning all the details in advance, which for Rocky that takes all the juice out of it and he becomes bored and wants to get out. He asks to go to the other place, hell, but his guide says he can't go there.

So he asks, "But I thought this was heaven?!"

And his guide says, "Whatever made you think this was heaven? This is Hell, Mr. Valentine! Ha, ha, ha"

The point being, that life is only really interesting if it is challenging, and what's more challenging than working your way all the way back to the level from whence we all started?

Ever watch a really interesting, serious movie and then suddenly it was written wrong or some dumb thing happens and it loses its edge? How did you react, with disappointment, right? Life is like that - it has an edge to it. We know we are mortal and aging, so we know we only have so much time to get 'it' right, 'It' being perceived differently for each of us. We push and get stressed out about things that really mean zero, zilch, nada in the bigger picture of how the Universe operates, yet we are crushed by those little disappointments. We live for it, we must have it. But that is also what makes life interesting. If suddenly we had every single thing we wanted completely satisfied like Rocky, then the edge, the challenge would be lost and we would have to reinvent ourselves to create a new edge, a new sense of 'Oh my God', I have to get this next thing right!

Not only does our consciousness inhabit a life form which is the result of millions of years of evolution, but we express ourselves as a fragmented lower version of God Consciousness. As such we are incomplete, and subject to all the complications that can occur when something perfect is reduced to its lesser derivations. In light of this, we may want to strive to be less judgmental and more understanding of ourselves and others. After-all, it is in this imperfect state in which we continue evolving, experiencing an individuality expressed in a uniquely creative way, that makes us so human. Sooner or later the train station is there for everyone, so for now I say ignore the final destination and enjoy your life in whatever pursuit drives your passions.

20.   Can you Paradigm Shift?

Historically, there has been polarized positioning between Science and Theology for the truth behind our Universe. A classic example of this dichotomy was a debate that ensued between these two institutions, initiated in 1927 when, Georges Lemaitre, a priest who taught physics at a Catholic University in Belgium, proposed in a Science article that our Universe was expanding. He made this leap from Einstein's Theory of Relativity, connecting the Universe together in a 4th dimension, which Einstein referred to as the fabric of space-time, (due to its interweaving of space and time). Lemaitre's position was that if the Universe is held together by a space-time fabric, then the Universe must be expanding or it would have collapsed long ago. Ironically, although Einstein's Theory does infer expansion, Einstein ignored the implications of his own Theory and stuck with the Steady State Theory, adhered to by most scientists of the time, suggesting all celestial bodies have always existed in their current orbital locations in the Universe.

Just two years later in 1929, Edwin Hubble discovered the Red Shift. An astronomical observation indicating galaxies are moving away proportional to their distance, meaning away from a point of origin, a Big Bang expansion. However, as great a discovery as this was, at the time it was not considered conclusive enough to quell the debate. In fact, cosmologist

Fred Hoyle found the idea of a Universe with a beginning to be philosophically troubling, as many including Hoyle argued that a beginning implies a cause, and thus a creator. The debate raged on until to two radio astronomers working for Bell Laboratories in 1965, accidentally picked up a background microwave signature of the Big Bang, proving it had occurred. The evidence was in the form of static sound from the expansion, but more importantly a previously predicted measurable temperature was verified, which would have only been detectable if a hot plasma Big Bang had occurred. At that juncture in the Scientific community, Steady State was out, replaced by the paradigm shifting knowledge gained of an expanding Universe. As you might have guessed though, this did not conclude the debate regarding its origins, with Science to this day holds to accidental reasons for the Big Bang, while Theology stands pat the Universe was God created.

Let us review for a moment some of the other paradigms that have occurred since that first person initiated a spark of light to control a fire. There was our realization of a spherical Earth. Then the paradigms started coming faster with the knowledge gained that life does not spring forth by way of Spontaneous Generation (e.g., frogs do not spring forth from water and mud, only frogs can make more frogs), the Earth is not the center of the Universe but neither is the Sun, Specie Evolution, the inner Universe too small to see with the naked eye, the fabric of space-time, our galaxy is not the only one - there are billions of galaxies, the Big Bang expansion, the Strong and Weak atomic forces and black holes.

As we can see from this progression, our collective perspective of the Universe has always presented itself on a grander scale with each successive shift in viewpoint. Yet, even with those many progressions there is a paradigm shift in our near future possessing such magnitude, it will cause a more dramatic alteration in human perspective than all those proceeding it put together a thousand times over.

This will transcend consciousness from humanity having every conceivable world view notion of the Universe, to a majority of people being on the same page to understand what caused the Big Bang and in so doing the design of the Universe. With that shift will infuse us with an understanding of our deep connection to all life, human and otherwise, of who we are, where we are going and how long we have. It will not happen overnight, and it will be fraught with outrage and dismissal by the extremists of

Science and Theology. Yet those with the strength to have maintained an open door to a fuller explanation of our wondrous Universe will be moved to strong emotions, as this new paradigm shift raises our individual and collective thought level to a whole new vantage point.

Science as an institution in pursuit of truths about our Universe, has uncovered so many mysteries that sometimes we are awe struck by its ability to explain naturally occurring phenomenon. As such, we presume it will answer all the remaining questions that can be asked, such as, 'What caused the Big Bang?', or 'What was the source of energy infused into the mass in the Big Bang?' However, in spite of proof in 1965 of the Big Bang, no answer has filled either one of those two biggest questions during the most technologically advanced period in human history.

Possibly life-forms get a late start to the party as mass compresses to denser elements, and the Universe continues to fully develop. Or maybe our lack of ability to astronomically view planets in the 'Goldilocks' distance from their suns has limited our view of how much life exists in our Galaxy and the Universe as a whole. In any case, it is likely there is a lot more consciousness in our Universe, because it exists in the 5th dimension, a dimension which we cannot directly detect. However, what is anticipated is experiments that test the first and fourth predictions of this Theory by independent sources to verify the validity of this theory.

In the spirit of Lamaitre, if you are geared primarily towards Science or Theology, cleanse yourself of any bias in either discipline in preparation for a Theory that combines both equally. For it is only in understanding the interdependence and interaction of both dimensions, that we can fully appreciate the full flowering of infinite possible third dimensional design and infinite possible consciousness perspectives.

I am certain, that during each Universe expansion, consciousness always reaches a point on each evolving planet, where upon the conceptual knowledge of how the Universe actually works is achieved. That is, the leading edge of consciousness clarifies an understanding of the Master Universe Design. And it will not be just Science or Theology, but rather a perfect blending based on equal amounts of matter in both realms, moving us towards a convergence of Science and Theology to more common ground. That time has arrived here near the beginning of the 21st Century.

21. Gemstones Found along a Circuitous Theoretical Path

While developing this Theory, there were numerous conceived ideas, yet the progression of each chapter had to proceed without necessarily expounding on every thought and thus some gems did not materialize into ink. But they are definitely worth perusing, and have therefore been collected together here in their own chapter.

- Upon leaving the theatre after watching the first of 'The Matrix' movies, my wife, Paula and I noticed ourselves and the other people exiting all felt elevated, as if we each possessed a new measure of thought power over our reality. Of course it was just a temporary state and quickly passed, yet that feeling represented a parallel in regards to thought level ascension. Although Neo lived in a program, the Matrix, which was fictional as compared to our reality as it pertained to what could be accomplished through the power of the mind, over the course of the movie we got to view him quickly transition from a person not too certain of his abilities, to having the power to stop a hail of bullets. I think this tapped into people's 'innate' understanding that we each have the power to transition to higher thought levels, and that someday in this life or in a future lifetime, we will each feel like Neo, having a much more powerful ability to influence our reality. Although we probably will not be able to stop a hail of bullets or physically move superfast, we will have much greater control over our life experiences in a positive manner.

I suggest the 2nd installment would have been better with Neo teaching the middle earth people how to do what he did, which near the conclusion of the first movie was insert himself into an agent and blast him into non-existence. Once all of them had attained the same thought level abilities as Neo, the ending scenes would have been a big party (as shown in the sequel), but up on the surface to celebrate ridding themselves of all the agents. Where did the message of ascending consciousness get lost?

 - Another sci-fi movie along these lines is 'Dark City'. In his search for Shell Beach, our star character, John Murdoch transitions to higher thought levels, eventually having the thought power to create a beach. The star of the movie, Dune, also makes the leap to higher thoughts, attaining the highest thought level possible in their fictional solar system. In each of these movies they ask if he could be the one, when in reality, at least as far as this theory professes, we all have the same capability to achieve the highest thought level possible.

 - The reason a god disburses 5th dimensional light, is for the opposite reason a black hole absorbs light, and that is because a god is at peaked energy and a black hole is at minimal energy. A god is peaked with energy, elevating the energy level of any and all incoming consciousness, and a black hole is at minimal energy, ejecting incoming energy in mass, including energy in light. If we could view a god, we would observe a bright white light. One way to accomplish this artificially would be to photograph a black hole and observe it in the negative.

 - The reason the black hole core mass must be reduced to minimal energy, indicative of its extremely cold temperature, is because it must maintain a uniform maximum density to remain in that most extreme compressed state, in which the atomic Weak force remains dominant, ready for the infusion of thermal energy it will receive to cause the Weak and Strong forces to switch roles, causing the instantaneous expansion of mass in the Big Bang.

 - One interesting perspective on this idea of all consciousness ascending into a composite God, is that all the consciousness evolution that takes place over the duration of the Universe by each of us individually, (as our collective thought at rest is released into the mass), is represented in the magnitude of the force infused into the Black Hole core to initiate the Big Bang. In this sense, we all take part in applying our ascended consciousness to this event, and in so doing take responsibility not only for the event itself, but for our own existence in the form we are now experiencing.

 - Both forms of matter's energy is incapable of reducing to zero, only minimal, i.e. a minor amount of energy is always present in mass and consciousness. The reason why, is the lowest form of consciousness disbursed in the Big Bang must have some energy, or else it would be incapable of fusing with earliest arising life-forms, and therefore due to the design parameter of a parallel Universe, mass cannot be reduced to zero energy either, only minimal. On a simpler level, we all understand that if a trickle of energy remains in a battery, it will recharge must faster than if it is reduced to zero energy. This concept carries over to both forms of matter in the Universe as well.

 - The existence of a similar component size to atoms (motas), and the Strong and Weak, dominant and recessive states in consciousness, are such that we probably cannot prove their existence, however if we can indirectly prove all other aspects of these parallels between the two forms of matter, then their existence is proven by solving for the unknown based on the other provable aspects of the overall design of the Universe.

 - By comparing mass & consciousness, we can see the majority of mass coalesces into black holes, and in turn we can presume the majority of consciousness coalesces into gods, versus ascending to that level. However, it is interesting to speculate as to whether by virtue of being at the human level at this fairly early stage in the duration of the Universe assures us that we will become Enlightened rather than coalesced.

 - At a certain point in the duration of the Universe, when mass has collectively depleted half of the energy originally infused into it in the Big Bang, and simultaneously consciousness as a collective measure has achieved half of the energy needed to reach the highest level, is the point when the energy fields for both realms are equal. It is referred to in this Theory as the 'Crossover Equilibrium'.

 - The argument of a Universe that expands forever versus one that ends in a Black Hole is a circular argument from the standpoint that it always comes back around to what caused the Big Bang. At minimum, a transfer of energy takes place from a sending source, to a receiving source, with the obvious inference that the Universe contains two forms of matter, each capable of containing energy. But more importantly if one of those forms of matter is consciousness, obviously if it is going to go to the trouble of Opposite Sub-Verses, opposite forms of matter, etc. then oscillation is certain.

- It does not take a lot of modeling of the Universe to realize that it must have two dimensions that each have a form of matter that holds energy, with one that is physical and the other must be non-physical. Two physical, or two non-physical dimensions as designs for the Universe, achieves nothing more than one of either. One physical loses energy over time with no way to charge the mass again, and one non-physical is insufficient, because there would be no physical dimension capable of supporting life forms to support consciousness evolution. Thus, the complexity of the design must be sufficient to achieve an infinite flow of energy, infinite forms of mass and consciousness, supportive of life forms, in an oscillating Universe, whereby mass and consciousness are cycled, which can only be achieved by one set of mirror imaged dimensions, that produce in the expansion an intermediary dimension, the 4th.

- Additionally, since no physical phenomenon exists that can penetrate the extreme density of the Black Hole core at the end of the Big Crunch, to infuse the thermal energy needed to cause its expansion, if a non-physical force, God, with peaked energy, does not exist to make the energy transfer, it would therefore not be possible for the Universe to renew itself, or to have caused the first ever Big Bang. This establishes a solid argument for equal amounts of opposite forms of matter.

- Although the Universe is expanding, is all mass orbiting the origination point of the Big Bang? The rate of rotation would be relatively slow, yet the question is of interest since all celestial bodies orbit the next largest gravitational source. The Earth orbits the Sun, and our solar system orbits a super-massive black hole at the center of our galaxy (Milky Way), but does the galaxy and the rest of the Universe orbit the origination point of the Big Bang?

- Someone could be brilliant intellectually, but if he or she is unable to connect at deeper levels with others, then their consciousness evolution is limited, until something changes to move the process forward. By the same token we can see that a dog can show unconditional love, but it's limited by its ability to gain knowledge. Thus, there is a trade-off depending on the state of a person's mind and or the life form the consciousness is infused into as to the extent of thought level advancement possible. Being human supports the highest potential intellect level of any life form on our planet, yet obviously presents a challenging medium at times with which to feel Love at deeper levels. Sometimes it's easy, sometimes it's not, which means we are at the right thought level for our current individual and collective progression.

"It's easy to love the whole world, but just try loving one other person", John Lennon

- How the need for deeper connectedness between people, which is mandatory for achieving higher thought levels, measures up to the increasing superficiality of ever greater and denser city populations, is an issue that will eventually need to be addressed. After all, there is little if any value to a society that reaches a point of pure production, void of meaningful interpersonal interaction, if it achieves little if any individual consciousness evolution. The question is; how do we balance our preoccupation with the multitude of demands in our lives, while also finding ways to create a civilization that is conducive to deeper connections?

One possible answer to this question may be the huge success of social connection websites. It makes it easy for likeminded people to connect.

- The idea that thoughts have power has been so subtle, that it has eluded detection in much the same way that Scientists once thought the Universe was static (early 20th century), then realizing later that it was expanding from a point of origin (Big Bang). However, this understanding that thoughts have power in consciousness is merely the eventual progression of cumulative thoughts arising over time from the many thousands of Scientists that have contributed towards this pinnacle of realization.

- Since our understanding of the non-physical is circumstantial to one degree or another beyond the obvious signature of the big expansion, we cannot judge whether our thoughts exist beyond the moment we conceive them, momentarily or permanently. Are they intertwining with all other thoughts in the non-physical 5th dimension, for any and all consciousness to conceivably draw from, or what is commonly referred to as the collective unconscious, as originally proposed by Carl Jung. However, scientific and inventive breakthroughs often come from different scientists that just happen to be studying and experimenting on the same topic simultaneously, reflective of the idea of a collective unconscious.

- Is our species highly cognitive enough to continue to consciously evolve, or will we cause a mass extinction, forcing the start of a new species that will continue the process? Ironically enough, in either case our individual consciousness will take advantage of whatever species is the brightest, infusing into it for our continued thought level ascent. But here are a couple of questions to ask yourself: In the case of an extinction of Humans, how long would we wait for a replacement species to infuse into, to continue our ascent to higher thought levels? But in that event, would we not expect a harsh repercussion for not being more responsible with the species we had previously utilized and the planet we occupied? Maybe that repercussion would take the form of an extended bout of boredom as we simply wait it out, and what is worse than boredom?

- In the aftermath of the Big Expansion, energized mass is disbursed into an expanding Universe, with lowest form minimal energy consciousness similarly being disbursed. It's not a bad thing though that consciousness is at its lowest level, because it will take several billion years before enough stars burn out and supernovae to cause the creation of remnant material forming planets that will support life, and super-massive black holes with a gravitational force that holds planets and stars in orbit around them to form galaxies, to create the necessary conditions for life to evolve. It is a long wait for life forms to start the physical evolutionary process, but time passes quickly in this state of minimal thought. Think of these thoughts as being extremely long, barely registering wavelengths. By the time our consciousness has evolved far enough to realize exactly how the Universe oscillates, presumably by way of this Theory, approximately 13.7 billion years will have passed.

- Was thinking through this Theory taxing? Believe me I lost a lot of sleep! In fact, during the latter part of the Theory development in which new ramifications, interactions, reconciling various concepts and other amazing thoughts were pouring in daily, I lost so much sleep I could have easily blown a gasket. Fortunately my mind and body were able to handle the stress and the Theory was completed. One might suggest, like my wife

Paula often did, to stop for a while. But how do you stop thoughts from cascading into your mind when you know if you do not capture them, they will be gone and you will have to start the whole momentum of thought progression over again? When you are in a multi-decade process of pushing the outside of the envelope to know the design of the Universe, you do not put up a stop sign or even a yield sign. You just hold on for dear life and take that thought roller coaster wherever it goes, from one side of the Universe to the other and back.

 - Hypothetically, if one were to walk out onto a beach stretching out in each direction as far as the eye can see, and toss a single grain of sand into the air, it would represent the Earth. Then, if all the other grains on that beach were blown up into the atmosphere in a strong wind, they would represent all the other celestial bodies of the Universe. With this image, we realize our experience individually and collectively on Earth is somewhat generic in relation to all the other planets that must be processing through the same Evolution, yet we each experience the full spectrum of consciousness evolution on our ascent to the highest thought level.

This Theory represents a bridge to answering the holy grail of scientific questions, 'What caused the Big Bang', answered quite simply by, 'We all did'. We all took part in the Thought that initiated the Big Bang, and we all experience a multitude of different life-form perspectives as we evolve to higher thought levels on our personal trek towards Enlightenment, but trying to get society as a whole to accept these ideas will probably be like the lyrics Edie Brickell sang in her song 'What I am'; "Religion is a smile on a dog" or "Philosophy is a walk on slippery rocks." Meaning, the trip there will have many interweaving chicanes, because the range of Human perspective on such thought provoking interpretations spans seemingly infinite possibilities and in most individuals is locked in like the gold at Fort Knox. That is why it is so important for there to be scientific proof to help move consciousness forward.

22. Paradigm Shifts, Past & Present

This chapter could have easily been placed near the beginning of the book however I did not want to lose readers that may not be as interested in Humankind's ascent. Yet it represents such a fascinating journey that it is highly recommended to help provide a historical perspective on the paradigms most of us will soon be passing through.

If you look closely at the human physique versus other predators, we are not well equipped to take down game or to make the kill once we have, and as such we probably started out as omnivorous foragers, a harsher term being scavengers. A million year old bone pile in Africa was uncovered, revealed shattered animal bones with chip marks on them from stones. Evidently early man used stones to crack open bones left by other predators to get out the marrow, a nutritious protein filled food source. In essence, due to our comparatively weak physical stature, all indications are that we were a prey specie living on the fringes. However, about 3.2 million years ago, an apelike ancestor stood up, yet still had the small brain pan of a lower primate. We know this from fossil remains of Lucy, the knee bones of which were oriented for an upright stance, yet with a comparatively small cranial cavity. And thus, it appears we first stood up, and in so doing freed our hands to do other things, such as tool making etc., leading to a larger brain to accommodate those new skills.

At some point we became predators, and over time almost all animals except a few became afraid of our specie. Recent footage taken in Africa showed three Bushmen of the Kalahari, dressed in traditional draping robes, as they made their way across the plains on foot holding long spears as walking sticks. Several hundred yards away, a lone full grown male lion caught sight of them and ran away as fast as his legs would carry him. This is a testament to our transition from prey to predator.

Once we discovered fire, we had essentially removed ourselves from the wild kingdom. Not only had we become predators, but we controlled something all other animals feared, and with our fear of the wild diminishing, this freed up more time and opportunity to increase our weapon making, cave art and communications skills.

The next major development was our control over agriculture. This undertaking got its start some 10,000 years ago and developed since then into the highly mechanized process we know today. But what was the psychological shift that took place by developing agricultural skills. Well, for one it was another separation from the wild kingdom around us, and this is important to look at because each time we separate ourselves, we in effect develop our specie into something unique that no longer has the constraints of a pure survival mode existence. Included in this time period is the beginning of our control over livestock. Agriculture and herding livestock assured a more predictable food source, providing more time for the arts and technological advances, and the start of written forms of communication.

And although technology did develop in these latest centuries, early in our history art played a much bigger role. The reason for this is because we didn't have the technology to make things fast and perfect each time. We had to rely on our dexterity and learned skills to make tools, weapons, baskets and pottery. Since everything was made by hand, the artisan had his or her own take on the final look, which usually included some adornment of color or unique shape. It is amazing to look at the ancient Egyptians and Romans from the perspective that their lives were still in a time period when everything was still done by hand, yet took on the finished look of temples and aqua ducts using arches, etc. Man was just beginning to make the transition where technology in design was changing from pure art, to art mixed with technology. It's interesting to note our fascination with that time period, because it was one of so much art and never again will our lives be filled with that level of artistry again. For everything gained something is lost. What's lost is a lifestyle where art is so much a part of life, but what is gained is that mass production allows for many more people to enjoy items of comfort and luxury.

Understandably, separating ourselves from the wild kingdom and the development of technology from a start of all things fashioned as art are important paradigm shifts that have taken place over thousands of years, and will continue to develop in the years to come. However, paradigm shifts that help us to understand the Universe and our place in it are the crux of where this is leading to, and will act as a base for the paradigm shifts of our present day.

Paradigms always start out as questions, as man asks the inevitable next important question in his pursuit of greater understanding. The first major paradigm was the question; 'What are we living on?', with the presumption that it is flat. It's not hard to grasp their mindset. After-all, unless one can travel far enough and fast enough, it would appear that it is a flat world due to its relative immensity in relation to our size, masking its overall sphere shape. Most people never ventured very far from their land, and so for them the explanation that the World was flat was readily acceptable. However, it is estimated that humankind figured out the world was a sphere about 2500 years ago, theorized to have occurred in part due to observations of an ocean's curved horizon.

In 500 BC the ancient Ionian philosopher Thales imagined the world was a giant disc floating on a cosmic sea of water. Anaximander said the world was curved, not flat, and pictured it as a giant cylinder surrounded by crystal spheres that held the Sun and the stars. A few years later in Athens, Greece, Democritus taught that all things were made of tiny particles he called 'atoms'. He also taught that all the specks of light in the Milky Way were stars just like our Sun, and that the Moon was an Earthlike body with mountains and deserts.

Born about the same time as Democritus died, was the most famous scientist of antiquity, Aristotle. He pictured the earth as spherical in the center of a series of fifty six concentric spheres. These spheres held the Sun, the Moon, the planets and the stars. The heaviest part of the Universe he said was the central part the Earth. That was why things fell down, he said. Next heaviest was water. Then came air and still lighter was fire which always rose up. And finally, highest and lightest of all were the crystal spheres of the sky, which held the Sun, Moon, planets and stars.

In the 2nd century AD, Ptolemy, a Greek astronomer simplified the picture first created by Aristotle and refined later by Greek astronomers. It was this picture of the famous Ptolemaic system that all educated people in the Western World held to throughout the Roman Empire and Middle Ages, a period of over one thousand five hundred years!

In 1534 Nicolas Copernicus, a Polish astronomer put forth the first real challenge to this system, developing a more accurate calendar to predict the position of the stars and planets. To do this Copernicus assumed the Sun was at the center and held that the Earth and other planets orbit around the Sun.   This made the motion of the planets easier to understand. In 1543 his book "On the Revolution of the Celestial Spheres" was used as a more accurate calendar for predicting the location of the planets however his view that the Sun as the center of the Universe, and not the Earth, was still very hard for people to accept.

A German astronomer, Johannes Keplar, took the next step by pointing to the need for elliptical, rather than circular orbits for the planets. He created the first laws of planetary motion showing how the planets in their elliptical paths would trace out equal areas in equal amounts of time.

Galileo Galilei, as a professor of astronomy at the University of Pisa, was required to teach the accepted theory of his time that the Sun and all the planets revolved around the Earth. Later at the University of Padua he was exposed to a new theory, proposed by Copernicus, that the Earth and all the other planets revolved around the Sun. Galileo's observations with his new telescope convinced him of the truth of Copernicus' Sun-centered or Heliocentric Theory. Unfortunately Galileo's ideas eventually brought him into conflict not just with the scholars of his day, who still believed in the authority of Aristotle, but also into serious conflict with the Catholic Church. In 1633 an inquisition convicted him to life imprisonment, but because of his advanced age he was allowed to serve out his term at home.

There was the realization that the heliocentric view was also not true in the sense that the Sun was not the center of the Universe, but one of innumerable stars. This was strongly advocated by the mystic Giordano

Bruno; Galileo made the same point, but said very little on the matter, perhaps not wishing to incur more of the church's wrath. Over the course of the 18th and 19th centuries, the status of the Sun as merely one star among many others became increasingly obvious. By the 20th century, even before the discovery that there are many galaxies, it was not an issue.

In the middle ages people believed in 'Spontaneous Generation', which was a misconstrued interpretation of their surroundings to believe that non-living objects could give rise to living organisms. For example, they believed muddy soil gave rise to frogs, mice came from moldy grain, sewage turned into rats, and flies came from dead meat. A recipe for bees was to kill a young bull, bury it upright so the horns protruded from the ground and after a month it will produce a swarm of bees.

In 1668, Francesco Redi conducted a scientific experiment by placing meat into a series of jars. He showed how the closed jar did not give rise to flies, but the uncovered jars did, concluding that only flies could make more flies. This experiment and others to follow finally dispelled the notion of Spontaneous Generation.

In 1839 Charles Darwin challenged humankind's view of himself with the Theory of Evolution. The public's response centered round their perception of Darwin's sheer audacity, the unmitigated gall to suggest humankind had descended from an 'apelike' ancestor. At the time it was a huge leap of belief and understanding. His theory was ridiculed in the extreme with articles and cartoons depicting man as an ape. However, his theory stated an 'apelike' ancestor, not an 'ape', but that image was used to try to discredit his Theory. Since then it has become common scientific knowledge from fossil records of other apelike hominids that roamed the Earth, that Homo sapiens are just one of many apelike species from the family tree of primates. For a variety of reasons all the other sub-species became extinct. In fact, DNA records show Homo sapiens (humans) went through a genetic bottleneck about 100,000 years ago, with all people today being descendants from just a few thousand people.

In any case, humankind learned from Darwin's theory that we evolved just like all the other animals by way of 'natural selection', or what is often referred to as 'survival of the fittest'. And since then the evidence put forth by scientists has clearly ratified his theory. In fact, humans have all the same genes as chimpanzees, except for about 2%.

There have been many major paradigms in Science, but one most people are aware of is the most famous equation, Einstein's E=MC2 first published in 1905, which explained in simple terms the power in the atom in mass 'at rest', by way of using the speed of light squared as a constant. This helped Scientists to understand that if they could release this energy, it would cause a huge explosion, namely an atomic bomb, or later with the hydrogen bomb. This discovery was not a paradigm from the standpoint that it caused a lot of debate amongst the average person, yet the information garnered from that equation represented a huge leap forward in our knowledge of the Universe in the field of Physics.

In the early part of the 20th century Georges Lemaitre, a priest who taught science at a Catholic University, suggested our Universe was expanding at a time when there was no knowledge of a Big Bang. He made this leap from Einstein's General Theory of Relativity, which connected the entire Universe together with a 4th dimension, Einstein referred to as the fabric of space-time, because it interweaved space and time. Lemaitre's position was that if the Universe is held together by this fabric so to speak, then it must be expanding or the Universe would have collapsed long ago. Ironically enough, although Einstein's Theory does infer expansion, Einstein ignored the implications of his own Theory and stuck with the 'Steady State' Theory, suggesting all celestial bodies are permanently in their current locations, which was the prevailing idea of the time.

This ongoing debate had scientific and theologian implications, from the standpoint that the Steady State position was strongly rooted in Science, whereas an expanding Universe had its implications with a God originated Universe. The debate raged on until to two radio astronomers in 1965, working for Bell Laboratories, accidentally picked up a background microwave signature, of the Big Bang, proving it had occurred. As Heidi Klum on Project Runway says, "Either you are in, or you are out", and The Steady State Theory was at that point, definitely out.

As you might imagine, this did not end the debate, because Science is still searching for physical reasons for the Big Bang while Theology simply stands pat that the Universe was God created. However, the next major paradigm shift will settle this argument once and for all, as we pursue answering a question we are apparently not suppose-to ask; 'What caused the Big Bang?

23. Thanks to Those that Contributed to This Theory

Where would this Theory be without all the contributions of knowledge provided by the numerous great scientists down through the centuries? Thus, below is a short list in recognition of some of the great scientists that contributed to the development of this General Theory. Also listed are some of the sources.

- To PBS for their first airing of 'The Origins of the Universe' in circa 1989, which lead to the initial hypothesis.

- To Nicolas Copernicus (Heliocentric Theory) and Galleili Galileo, for their combined efforts in helping to convince a skeptical world that the Earth is not the center of the solar system.

- Charles Darwin for his paradigm shifting Theory of Evolution, suggesting that humankind evolved from an 'apelike' ancestor, which has since been proven by fossilized remains and genetics. His Theory led to the symbiotic positive loop exchange, i.e. two interwoven layers of evolution.

- Albert Einstein for his equation $E=MC2$ (1905), which was critically important in the development of the equation $t=coC2$, for leading us to understand the 4th dimension fabric of space-time leading to my Theory's explanation for how it was generated out of the Big Bang, and for his statement that the correct understanding of how the Universe worked when it would someday be presented would be easy to understand, because out of simplicity arises complexity.

- Georges Lamaitre for his Theory in 1927 that the Universe was expanding, initiating debates between the theories of Steady State and Expansion, with Expansion winning out.

- Edwin Hubble for Hubble's Law in 1929, which showed galaxies are moving away at speeds proportional to their distance, a strong indicator of the Big Bang origin of the Universe.

- Sir Fred Hoyle, for his anthropic observation from the triple-alpha process that produces carbon, and once realizing just how specific the energy that was needed for producing carbon (necessary for life), for his strength of character in switching from an atheist to an agnostic theist.

- Arno Penzias and Robert Wilson, radio astronomers, who discovered a 2.725 degree Kelvin cosmic microwave background radiation, which is the remnant heat signature scientists were looking for to help prove the Big Bang event.

- Carl Jung, for his theoretical ideas regarding the collective unconscious.

- Brandon Carter, astrophysicist and cosmologist from Cambridge University, for his work on the Anthropic Principle, first suggested in a paper in 1973.

- 2004 Nobel winning scientists David Gross, David Politzer and Frank Wilczek, for their theoretical discovery that the strong force is the dominant one in the nucleus of an atom, which helped lead to the Atomic & Motaic Force Role Switching Mechanism.

- To Dr. Duncan MacDougall, for the WLD tests he conducted in 1907, that for their time caused quite a sensation, and which provided insight into developing my own methodology for the Consciousness Predictions. His hope was that although his test results did not prove the subject, it would serve to inspire more testing, which we can all hope does take place. His pioneering work stands as a testament to the human spirit of seeking greater knowledge.

- My Father Graham Slater, for his form of mechanical engineering logic ingrained into my way of thinking, that acted as the linear counter balance to the random manner in which creative ideas were intermixed by way of the autistic part.

- My Brother, Gary, for his independent way of conducting himself, which transferred itself to my way of doing things later in life, particularly in the development of this Theory.

- To my wife Paula Slater without whom this theory would not have come to fruition due to her loving willingness to let me be me, and I have done the same as her career has risen to becoming a great sculptor, since we first met in 87 and married in 88. See paulaslater.com to see her many completed projects.

- Any and all other people or scientists not mentioned which I am certain there are many, yet obviously contributed in many ways.

24. The Development of an Unusual Way of Thinking

(Written in the format of a movie script)

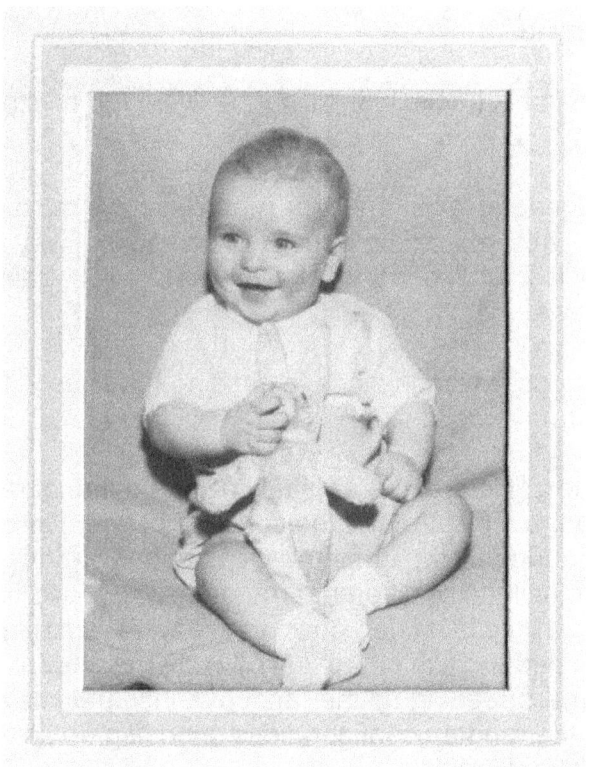

C. Nevell Slater at 7 months, Feb. 1957 in England

1959, Toronto, Canada. Move view to the suburbs, brick walled home with lawn out back. Moving inside there is Copenhagen furniture, with a blond boy 3-4 years of age running from the living room into the kitchen, through the dining area, back into the living room.

Using a voice over by Morgan Freeman; "As this young boy of three races around the bottom floor of their home, he is headed for a life changing event that would normally leave a boy of his age in an autistic state for the rest of his life. You see, his consciousness has awoken to this rich new environment, yet his energy keeps getting spent physically, not mentally. So his mind has not developed in thought level balance with his consciousness which could potentially cause his consciousness to partially dislodge from his mind."

(Nevell stops dead in his tracks just inside the living room with his eyes wide open.)

"His body can no longer get out the energy stored in his mind. But as he brushes up against this impending transition, he seeks a way to use the energy in one final last ditch effort to survive what he is certain will in some manner take him away from his Family."

(Show Nevell with legs frozen in place, extending his arms in front of him, moving his hands from side to side.)

Voice over continues: "He creates a visual imaging game of rearranging two items in the room, a picture and the couch to different locations in the room, then reconfiguring all the furniture, pictures and various items in the living room into a multitude of different arrangements. Each time reconfiguring them a little bit faster, down to the arrangement of the items on the coffee table and the fish in the tank. Like a maestro at a concerto he now owns this game, speeding up the random configurations at blinding speed." (Showing the random configurations speeding up as the voice over explains what is happening.)

(His Mother wondering why things have gotten so quiet leans around the corner by the dining area to see how he's doing and asks in an English accent.)

Marion: "Nevell, are you alright?"

(Nevell looks down at his legs and watches them as they can now move again, albeit deliberately one at a time at first.)

Nevell: "Yeah Mom. I was gonna hafta go away, but I won, so I can stay!!!"

(Voice-over continues: "He's successfully averted the full malady of Autism by increasing his brain's thought level to match his consciousness. However the way it was done changed his way of thinking to that of a random dyslexic").

(While drying a dish, she peers at him wondering if he is really all right with the doorbell ringing and the sound- segue to a school bell ringing, with Nevell in first grade in a Canadian school, in a suit and tie. He's standing up by his desk reciting the numbers one through ten.)

Nevell in Toronto, first grade

"One, two, three, four, five, seven, six, eight, nine, ten."

The teacher says wrong, and laughter rains down from the other students.

Teacher: "Was it so hard? I gave everyone a whole week to memorize and recite the count from one to ten, and Nevell you are the only one to fail! Why? From now on when you make a silly mistake like that I'm going to have to discipline you with this ruler (as she raises it up). Now I don't want to have to do that, so pay attention and keep up with the other students!"

(As she finishes her statement, the camera closes in on her mouth transitioning into full darkness-transitions into letters on the chalkboard cascading down like water falling at Niagara Falls, as Nevell is imagining this with the letters breaking apart into a mist of little black bits, spilling off the chalkboard shelf.)

Teacher: "Nevell, it's your turn to recite the alphabet."

Nevell: "Pardon?"

Teacher: "My goodness, you really need to pay better attention?! Now recite the alphabet."

Nevell: "A-B-C-D-E-F-G-K-N-P"

Teacher: "Stop. We will keep doing this until you get it right. ABCDEFGHIJKL"

Nevell: "A-B-C-D-E-F-J"

Teacher: "I will keep correcting you until you can recite the whole alphabet in the right order."

(Move the camera away, run some background music that is negative in mood, and show that Nevell continues to fail and the teacher continues to correct him multiple times. Show the other kids laughing and giggling.)

Nevell: "I can't do it."

Teacher: "Yes you can, now from the beginning once more."

(Show Nevell's frustration and him not answering but thinking instead.)

Nevell: "No."

Teacher: "Come on, it's easy. ABCD..."

Nevell: "TBSIOELCHAVPLZ... (Laughter breaks out from the class)

Teacher: "Stop! I will not allow you to make a mockery of this class - do you hear me!!"

(She gets up from her desk, grabs the ruler, raises it high above her head and brings it down over the back of Nevell's hand, with the sound reverberating off the walls. The other students wail and shriek with excitement and joy, laughing and jeering at him. Having lost her temper, the teacher says)

Teacher: "You are by far the worst student in all the classes I've ever taught! Not one time in any exercise have you shown any aptitude above the bottom of the class! You will begin to pay better attention and do better in this class, or suffer the wrath of this ruler!"

(She waves the ruler back and forth. As the teacher is talking, Nevell is inspecting his hand which is glowing red and white - he can clearly see bones on the back of his hand from the blow - which fascinates him, and the shot of his hand-to the back of Nevell's blond head of hair when he was in 3rd grade, then zoom back to bring into focus the special reading class he's attending. It has a long table, small room, 7 students in the 3rd grade. School design is simple art deco, still there in Sausalito, CA but used for city offices and a library.)

Nevell: "Jane says to Bill, throw the ...." (He stops reading because a random word on the page lit up in his mind, instead of the word he needed to read. The film needs to show the words lighting up as he reads, including the randomly lit word, so the viewer understands his condition, random dyslexia, as the reason he stops reading.)

Teacher's Asst.: "Just read the next word. It's all right we'll come back to you on Monday." (Said as she stands, with the kids also standing to leave)

Redhead girl: "When will we be going back to our regular class?"

Teacher's Asst.: "When you can read faster and without hesitation."

(All 7 kids: "AHH!" A sound they make due to being embarrassed by their need to attend this special reading class for slow readers. As this scene ends, the camera viewpoint should make it appear it is dropping down, like it's dropping through the floor into the room below, and as it does it-

    -transitions to the salmon color of the carpeting in Nevell's bedroom on the weekend after the reading class' previous scene. Nevell is sitting on the edge of his bed reading aloud alone, but he is only being able to read six or seven words in a row before stopping because his mind wants to read a randomized word. After each time he stops, he hits himself in the side of the head trying to will himself to be able to read straight. This is repeated three times then loses his temper and pounds the side of his head some dozen times in a row. Switch back and forth between showing Nevell in his room and his Mother outside the room, with her picking up on the sounds. His Mother comes in search of what is making the thudding sounds and as she steps into the room, she catches a glimpse of him pounding his skull.)

Mother (Marion): "I was wondering what that sound was Nevell, why were you hitting yourself?!!"

Nevell: "I can't read straight!"

Marion: "My, you're really having trouble reading aren't you? Do you want Gary (his Brother) to help you? Remember at one time he helped you to count to ten?"

Nevell: "No, I'll find a way!"

Marion: "Alright, but you have to promise me you'll never hit yourself again. Promise me now!"

Nevell: "Ok, I promise not to hit myself again."

(After she leaves the room, the camera needs to show the book pages, and as he reads aloud to himself the words light up straight across. He reads without stopping and starts to smile as he realizes he no longer has the same problem. His learning disability is suddenly gone).

Nevell: "I wonder if I can still do that random thing, but I don't want to do it with words. I'll try it with a picture."

Voice over by Morgan Freeman: "He wants to test his ability to randomly configure objects, but doesn't want to risk it with words, so he uses a picture in the book."

(Camera zooms in to show a page of the book with a picture of a boy, a girl and a dog. Use computer animations of the three of them as they randomly skip back and forth past one another up a small grassy hill, singing a tune with the dog jumping up to catch a ball in between them.)

Nevell: "Hey, I've still got it! I can read straight across, but I can also randomly rearrange things!"

Voice over continues: "Nevell has now come full circle from his initial brush with autism, successfully overcoming his learning disability, but also forcing a split to think in two different ways; in a conventional linear way or randomly testing different possibilities. This leads to him as an adult being able to conduct conceptual thought experiments, leading to the development of a revolutionary new theory explaining the design of the universe as Opposite Sub-Verses, The Engine of Oscillation."

For those wanting to contact me with positive comments or with information on efforts and experiments to assist in proving some of the ideas pertaining to this theory, please email:

oppositesubverses@gmx.com